Surface Chemistry Essentials

Surface Chemistry Essentials

K. S. Birdi

CRC Press is an imprint of the
Taylor & Francis Group, an **informa** business

CRC Press
Taylor & Francis Group
6000 Broken Sound Parkway NW, Suite 300
Boca Raton, FL 33487-2742

Printed on acid-free paper
Version Date: 20130925

International Standard Book Number-13: 978-1-4398-7178-2 (Paperback)

Library of Congress Cataloging-in-Publication Data

Birdi, K. S., 1934- author.
Surface chemistry essentials / K.S. Birdi.
pages cm
Includes bibliographical references and index.
ISBN 978-1-4398-7178-2 (pbk.)
1. Surface chemistry. I. Title.

QD506.B57 2014
541'.33--dc23 2013036998

Visit the Taylor & Francis Web site at
http://www.taylorandfrancis.com

and the CRC Press Web site at
http://www.crcpress.com

Contents

Preface

The subject related to surface chemistry is recognized as being an important area of a special branch of chemistry in everyday life. Due to the extensive literature devoted to this subject, it was realized that a book comprising only *essential* coverage would be useful. Thus, the aim of this book is to present and describe the essential details of surface chemistry. This need also arises from the fact that the area of applications in industry of this science is now extensive.

Most science students are taught physicochemical principles pertaining to gases, liquids, and solids. The matter around us is recognized to be made of these three states of matter. However, in university chemistry textbooks, only a chapter or two is devoted to the science of surface and colloid chemistry. At technical schools the case is the same; in general, it is the case worldwide. The science of surface and colloid chemistry is one of the most important in technology. The most common examples include:

Soap bubbles
Foam (fire fighting)
Raindrops
Combustion engines
Food products
Air pollution (fog, smog, sandstorms)
Wastewater treatment
Washing and cleaning
Corrosion
Cosmetics
Paint and printing; adhesion; friction
Oil and gas production, and shale oil/gas recovery (fracking process)
Oil spills
Plastics and polymers
Biology and pharmaceuticals
Milk products (milk, cheese)
Cement
Adhesives
Coal (coal slurry transport)

Science students are increasingly interested in the application studies to real-world systems. Colloid and surface chemistry offers many opportunities to apply this knowledge to understanding everyday and industrial examples.

The main purpose of this book is to guide chemistry and physics students with backgrounds in the area to the level where they are able to understand many natural phenomena and industrial processes, and are able to widen their application potential to new areas of research. The text is carefully arranged such that much involved mathematical treatment of this subject is mostly given as references. However, students should be able to do this and still maintain a good understanding of the fundamental principles involved.

Furthermore, this book contains useful data from real-world examples, which helps to explain and stimulate the reader to consider both fundamental theory and industrial applications. The latter is expanding rapidly and every decade brings new application areas in this science. Accordingly, pertinent references are provided for the more advanced students and scientists from other fields (such as biology, geology, the pharmaceutical industry, medical science, astronomy, and plastics).

Important sample questions and answers are included wherever appropriate in various sections, with detailed data and discussion. Although the text is primarily aimed at students, researchers will also find some topics of interest. A general high school background in chemistry or physics is all that is required to follow the main theme in this book.

During the past decades, it has become more obvious that students and scientists of chemistry and engineering should have some understanding of surface and colloid chemistry. The textbooks on physical chemistry do introduce this subject, but there is generally only a short chapter. Modern nanotechnology is another area where the role of surface chemistry is important. The medical diagnostic applications are another area, where both microscale and surface reactions are determined by different aspects of surface and colloid chemical principles. Drug delivery is mostly based on lipid vesicles (self-assembly structure), which are stabilized by various surface forces.

The book presents essentials and some basic considerations with respect to liquid and solid surfaces. After this introduction, the liquid–solid interface phenomenon is described. Following this, the colloid chemistry systems are discussed. The essential principles of emulsion science and technology aspects are presented. In the last chapter, more complex application examples are described. These are examples where different concepts of surface and colloid chemistry are involved in some mixed manner.

1 Introduction to Surface Chemistry Essentials

1.1 WHAT IS SURFACE CHEMISTRY?

Science is concerned with knowledge of the structures of matter (defined as solids, liquids, and gases) (Figure 1.1). Modern technology has shown that one needs a much more detailed picture of these structures in all kinds of processes (chemical industry and technology and natural biological phenomena). The modern application industry of science is clearly the most important area for mankind's demands (with regard to future challenges: drinking water, energy resources, food, clean air, transportation, housing, health and medicine, etc.).

Matter exists as:

- gas,
- liquid, and
- solid,

as has been recognized by classical science:
Solid phase → Liquid phase → Gas phase

Experiments show that molecules that are situated at the *interfaces* (e.g., between gas–liquid, gas–solid, liquid–solid, $liquid_1$–$liquid_2$, $solid_1$–$solid_2$) are known to behave differently (Figure 1.2) than those in the bulk phase (Bakker, 1928; Adam, 1930; Bancroft, 1932; Partington, 1951; Harkins, 1952; Davies and Rideal, 1963; Defay et al., 1966; Gaines, 1966; Matijevic, 1969; Aveyard and Hayden, 1973; Fendler and Fendler, 1975; Chattoraj and Birdi, 1984; Birdi, 1989, 1997, 2002CD, 1999, 2002, 2009, 2010a, 2010b; Adamson and Gast, 1997; Rosen, 2004; Schramm, 2005; Somasundaran, 2006; Kolasinski, 2008; Miller and Neogi, 2008; Somarajai and Li, 2010; Barnes, 2011). Typical examples are:

Liquid surfaces
  Surfaces of oceans, lakes, and rivers
  Lung surface, biological cells surfaces

Solid surfaces
  Road surfaces (car tire)
  Adhesion, glues, tapes

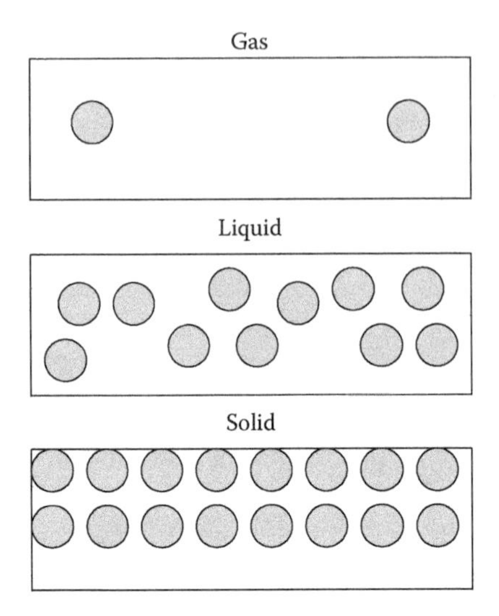

FIGURE 1.1 The molecular structure of a gas, liquid, and solid.

- Cement industry
- Paper industry
- Construction industry (tunnels, etc.)

Liquid–solid interfaces

- Washing and cleaning (dry cleaning)
- Wastewater treatment
- Air pollution
- Power plants

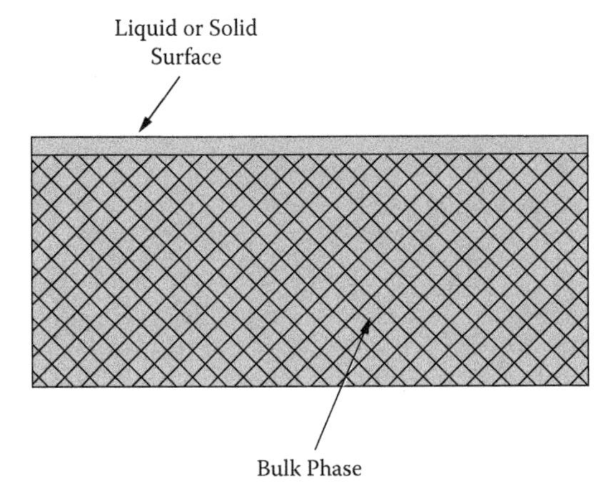

FIGURE 1.2 Bulk phase (liquid or solid) and surface phase (molecular dimension).

Liquid–liquid interfaces (oil–water systems)
Emulsions (cosmetics, pharmaceutical products)
Diverse industries
Oil and gas, and shale oil recovery (fracking technology), paper and printing, milk products

1.2 ESSENTIAL SURFACE CHEMISTRY CONCEPTS

The science of surface chemistry covers a very large area, and therefore some essential concepts are delineated here. More details will be covered in the rest of the book, and real practical examples will be analyzed. The classical physical chemistry will be applied throughout the book, along with suitable literature references. However, some essential principles will be delineated in appendices in each chapter.

As a typical example, reactions taking place at the surface of oceans (such as solubility of oxygen, CO_2, etc.) will be expected to be different than those observed inside the seawater. Further, in some instances, such as oil spills, one can easily realize the importance of the role of surface of oceans (Figure 1.3). It has been found that part of oil evaporates, while some sinks to the bottom, and the main part remains floating on the surface of water. This process is one of the major areas of surface chemistry applications. It is also obvious that the surface of oceans plays an important role in everyday life.

It is also well known that the molecules situated near or at the interface (i.e., liquid–gas) will interact differently with respect to each other than the molecules in the bulk phase (Figure 1.4). The essential aspects of this important subject will be described extensively in this book. The intramolecular forces acting would thus be different in these two cases. In other words, all processes occurring near any interface will be dependent on these molecular orientations and interactions. Furthermore, it has been pointed out that, for a dense fluid, the repulsive forces

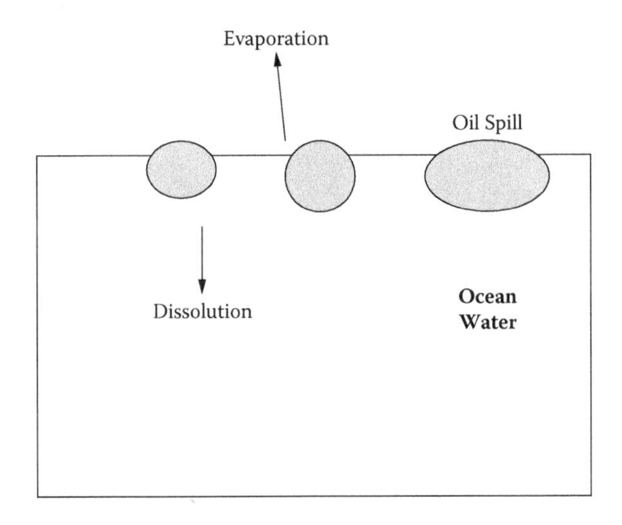

FIGURE 1.3 Ocean surface and oil spill (evaporation, solution, sinking, floating states).

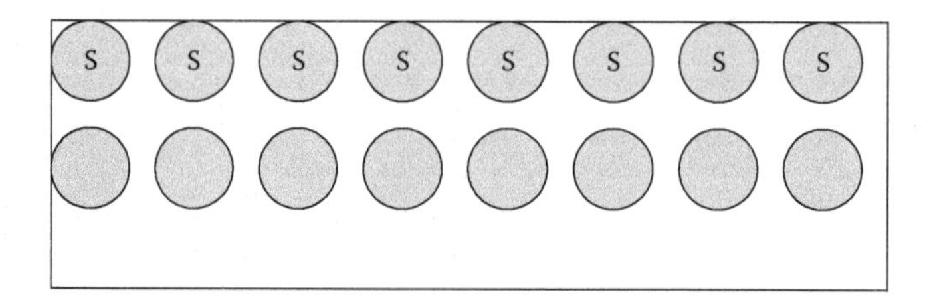

FIGURE 1.4 Surface molecules.

dominate the fluid structure and are of primary importance. The main effect of the repulsive forces is to provide a uniform background potential in which the molecules move as hard spheres. The molecules at the interface would be under an *asymmetrical* force field, which gives rise to the so-called surface tension or interfacial tension (Figure 1.4) (Chattoraj and Birdi, 1984; Birdi, 1989, 1997, 1999, 2002; Adamson and Gast, 1997).

At a molecular level, when one moves from one phase to another, that is across an interface, this leads to adhesion forces between liquids and solids, which is a major application area of surface and colloid science (Figure 1.5).

The resultant force on molecules will vary with time because of the movement of the molecules; the molecules at the surface will be pointed downward into the bulk phase. The nearer the molecule is to the surface, the greater the magnitude of the force due to *asymmetry*. The region of asymmetry plays a very important role. Thus, when the surface area of a liquid is increased, some molecules must move from the interior of the continuous phase to the interface. Surface tension of a liquid is the force acting normal to the surface per unit length of the interface, thus tending to decrease the surface area. The molecules in the liquid phase are surrounded by neighboring molecules and these interact with one another in a symmetrical way.

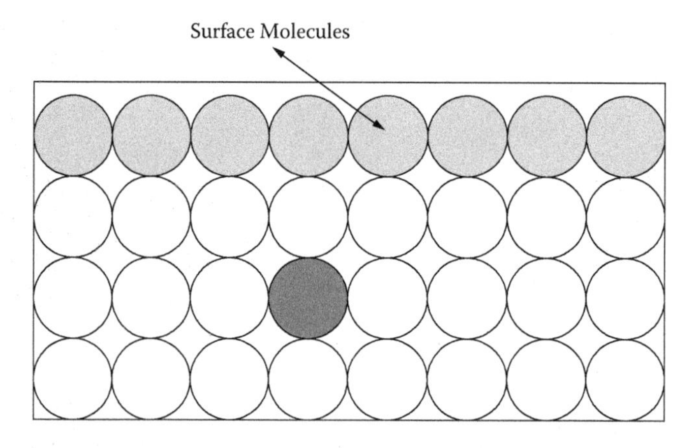

FIGURE 1.5 Intermolecular forces around a molecule in the bulk liquid (dark) and around a molecule on the surface (light) layer.

The state of solid–liquid–gas phases may be described in a simpler way as follows. These phases may be compared with a football field with people as atoms (or molecules):

- There are people sitting in their seats (solid phase).
- There are people moving in and out of their seats (liquid phase).
- There are players running around in the field (gas phase).

Note that the distance between players on the field is much larger and variable (even including collisions, exactly what also happens in the gas molecules) than those sitting in their seats, which is analogous to molecules in a gas.

In the gas phase, where the density is 1000 times lesser than in the liquid phase, the interactions between molecules are very weak as compared to in the dense liquid phase. Thus when one crosses the line from the liquid phase to the gas phase, there is a change in density of factor 1000. This means that while in liquid phase a molecule occupies a volume that is 1000 times smaller than when in the gas phase. The interfacial region is found to be of molecular dimension. Some experiments show it to be of one or few molecules thick.

Surface tension is the differential change of free energy with change of surface area. An increase in surface area requires that molecules from the bulk phase are brought to the surface phase. The same is valid when there are two fluids or a solid–liquid; it is usually designated *interfacial tension*. A molecule of a liquid attracts the molecules that surround it and in turn it is attracted by them (Figure 1.5). For the molecules that are inside a liquid, the resultant of all these forces is neutral and all of them are in equilibrium by reacting with each other. When these molecules are on the surface, they are attracted by the molecules below and by the lateral ones, but not toward the outside. The result is a force directed inside the liquid. In turn, the cohesion among the molecules supplies a force tangential to the surface. So, a fluid surface behaves like an elastic membrane that wraps and compresses the liquid below the surface molecules. The surface tension expresses the force with which the surface molecules attract each other. It is common observation that due to the surface tension it takes some effort for some bugs to climb out of the water in lakes. On the contrary, other insects, such as marsh treaders and water striders, exploit the surface tension to skate on the water without sinking (Figure 1.6).

Insects that move about on the surfaces of lakes are actually also collecting food from the surface of the water. Another well-known example is the floating of a metal needle (or any object heavier than water) on the surface of water (Figure 1.6). The surface of a liquid under tension maintains a sort of skin-like structure. In other words, energy is required to carry any object from the air through the surface of a liquid. The surface of a liquid can thus be regarded as the plane of potential energy. It may be assumed that the surface of a liquid behaves as a membrane (at a molecular scale) that stretches across and needs to be broken in order to penetrate. One

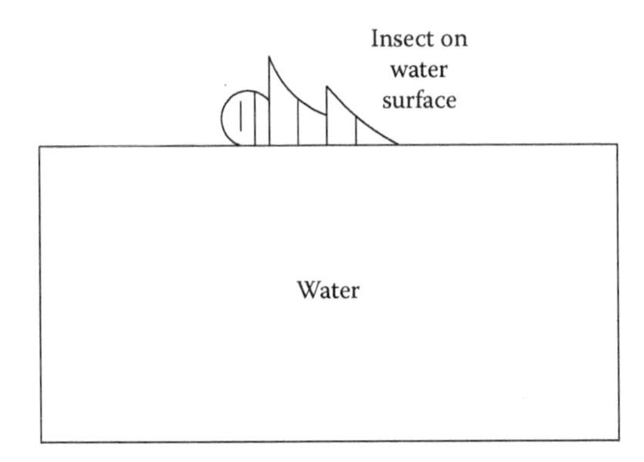

FIGURE 1.6 An insect (many different kinds of insects, including mosquitoes) strides on the surface of water.

observes this tension when considering that a heavy iron needle (heavier than water) can be made to float on a water surface when carefully placed (Figure 1.7).

The reason a heavy object can float on water is due to the fact that in order for it to sink it must overcome the surface forces. Of course, if one merely drops the metal object it will overcome the surface tension force and sink. This clearly shows that at any liquid surface a tension exists (surface tension) that needs to be broken when any contact is made between the liquid surface and the material (here, the metal needle). There are ample examples on the surfaces of rivers and lakes, where stuff is seen floating about. Based on the same principles, it has been found that the smooth hull of a ship exerts less resistance to sail than a rough bottom, thus saving energy.

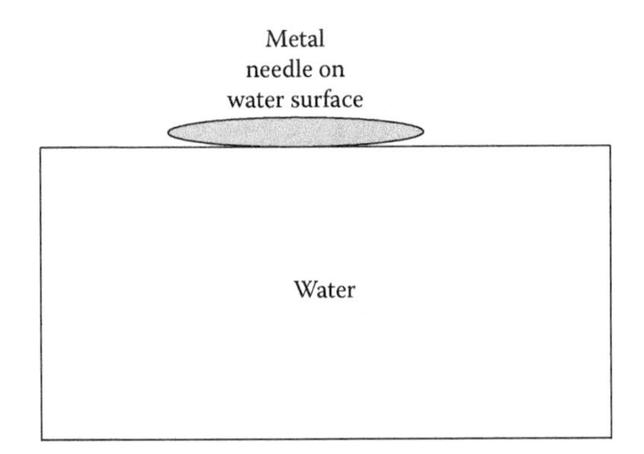

FIGURE 1.7 An iron needle (or any similar metal object) floats on the surface of water (only if carefully placed, otherwise it should sink due to gravity forces). The best procedure is to place the metal object on a piece of paper, and then place the paper on the surface of the water. As the paper sinks, the metal object remains floating.

Definition of liquid interfaces:

1. Liquid and vapor or gas (for example, ocean surface and air)
2. $Liquid_1$ and $liquid_2$ immiscible (water–oil, emulsion)
3. Liquid and solid interface (water drop resting on a solid, wetting, cleaning of surfaces, adhesion)

Definition of solid surfaces or interfaces:

1. $Solid_1$–$solid_2$ (cement, adhesives)

An analogous case would be when the solid is crushed and the surface area increases per unit gram (Figure 1.8). For example, finely divided talcum powder has a surface area of 10 m^2 per gram. Active charcoal exhibits surface areas corresponding to over 1000 m^2 per gram. This is an appreciable quantity and its consequence will be shown later. Qualitatively, it should be noted that work has to be put into the system when the surface area increases (both for liquids or solids or any other interface). Cement is mainly based on the energy used to make the particles as small as possible, such that the cost is dependent on this process.

The surface chemistry of small particles is an important part of everyday life (such as, dust, talcum powder, sand, raindrops, and emissions). Let us define what is meant by *colloid chemistry* and its relation to surface chemistry. This was already defined a century ago by Thomas Graham. A particle having dimensions in the range of 10^{-9} m (10 A) to 10^{-6} m (1 μm) was considered to be colloidal. The nature and relevance of colloids is one of the main current research topics (Birdi, 2002, 2010a). Colloids are an important class of materials, intermediate between bulk and molecularly dispersed systems. The colloid particles may be spherical, but in some cases one dimension can be much larger than the other two (as in a needlelike shape). The size of particles also determines whether they can be seen by the naked eye. Colloids are not visible to the naked eye or under an ordinary optical microscope.

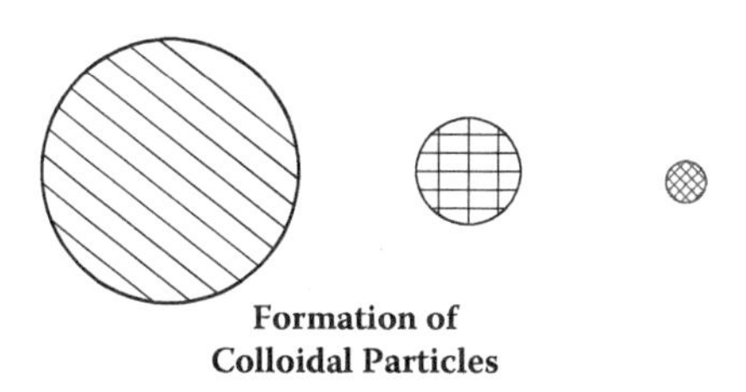

FIGURE 1.8 Formation of fine (colloidal) particles (such as talcum powder, active charcoal, or cement). (The size is less than a micrometer.)

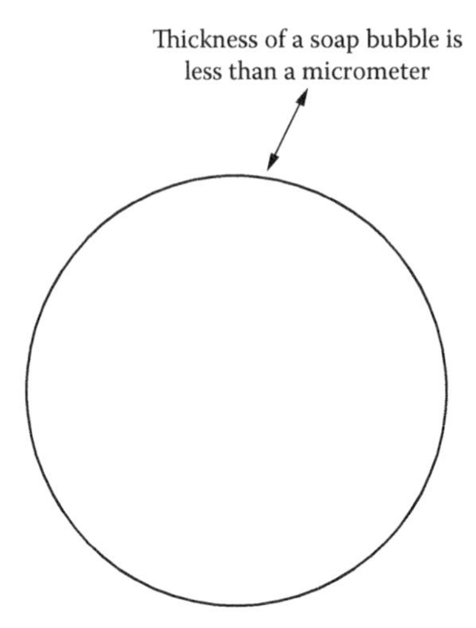

FIGURE 1.9 Soap bubble of thickness (micrometer or less).

The scattering of light can be suitably used to see such colloidal particles (such as dust particles). The size of colloidal particles then may range from 10^{-4} to 10^{-7} cm. The units used are as follows:

$1\ \mu m = 10^{-6}$ m
1 nm $= 10^{-9}$ m
1 Å (Angstrom) $= 10^{-8}$ cm $= 0.1$ nm $= 10^{-10}$ m

Nanodimension can be imagined by considering the following simple examples: Hair: 1/1000th diameter is about nanosize. The thickness of soap bubbles varies from micro- to nanometer colored rings (in Figure 1.9). Actually, a soap bubble is the closest thing that can be seen by the naked eye that is of molecular dimension.

In surface chemistry there is a great need for suitable range of dimensions as needed for a variety of systems. As seen here, the range of dimensions is manyfold. Accordingly, a unit Angstrom (Å $= 10^{-8}$ cm) was used for systems of molecular dimension (famous Swedish scientist). However, the most common unit is the nanometer (10^{-9} m), which is mainly used for molecular scale features. In recent years, nanosize (nanometer range) particles are of much interest in different applied science systems (*nano* from Greek and means "dwarf"). Nanotechnology is actually strongly getting a boost from the last decade of innovation, as reported in the surface and colloid literature (Rao, 2011). In fact, light scattering is generally used to study the size and size distribution of such systems. Since colloidal systems consist of two or more phases and components, the interfacial area to volume ratio becomes very significant. Colloidal particles have a high surface area to volume ratio compared with bulk materials. A significant proportion of the colloidal molecules lie within, or close to, the interfacial region. Hence, the interfacial region has significant control over the

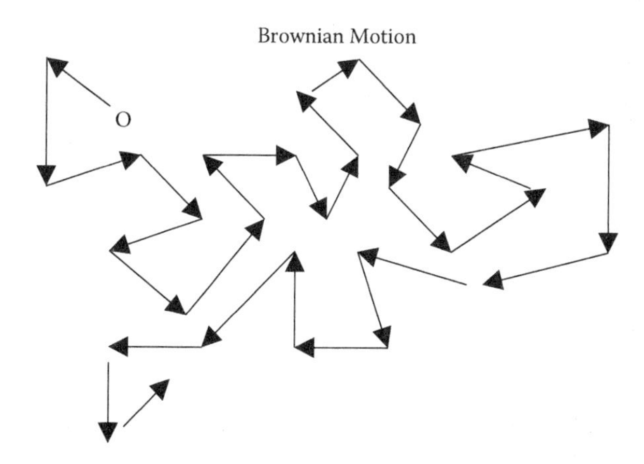

FIGURE 1.10 Brownian motion. Dust particles in the air are sometimes observed moving in jumps. This is due to molecular collisions between the gas molecules and the dust particles.

properties of colloids. To understand why colloidal dispersions can either be stable or unstable we need to consider the following:

1. The effect of the large surface area to volume ratio (for example, 1000 m^2 surface area per gram of solid [active charcoal])
2. The forces operating between the colloidal particles
3. Surface charges are very important characteristics of such systems.

There are some very special characteristics that must be considered about colloidal particle behavior: size and shape, surface area, and surface charge density. The *Brownian motion* of the particles is a much-studied field (Figure 1.10). The *fractal* nature of surface roughness has recently been shown to be of importance (Birdi, 1993). Recent applications have been reported where *nanocolloids* have been employed. It is thus found that some terms are needed to be defined at this stage. The definitions generally employed are as follows. *Surface* is a term used when one considers the dividing phase between:

- Gas–liquid
- Gas–solid

Interface (Figure 1.11) is the term used when one considers the dividing phase:

- Solid–liquid (colloids)
- $Liquid_1$–$liquid_2$ (oil–water, emulsion)
- $Solid_1$–$solid_2$ (adhesion, glue, cement)

It is obvious that surface tension may arise due to a degree of unsaturation of the bonds that occurs when a molecule resides at the surface and not in bulk. The term *surface tension* is used for solid–vapor or liquid–vapor interfaces. The term *interfacial tension* is more generally used for the interface between two liquids (oil–water), two solids, or a liquid and solid. It is, of course, obvious that in a one-component

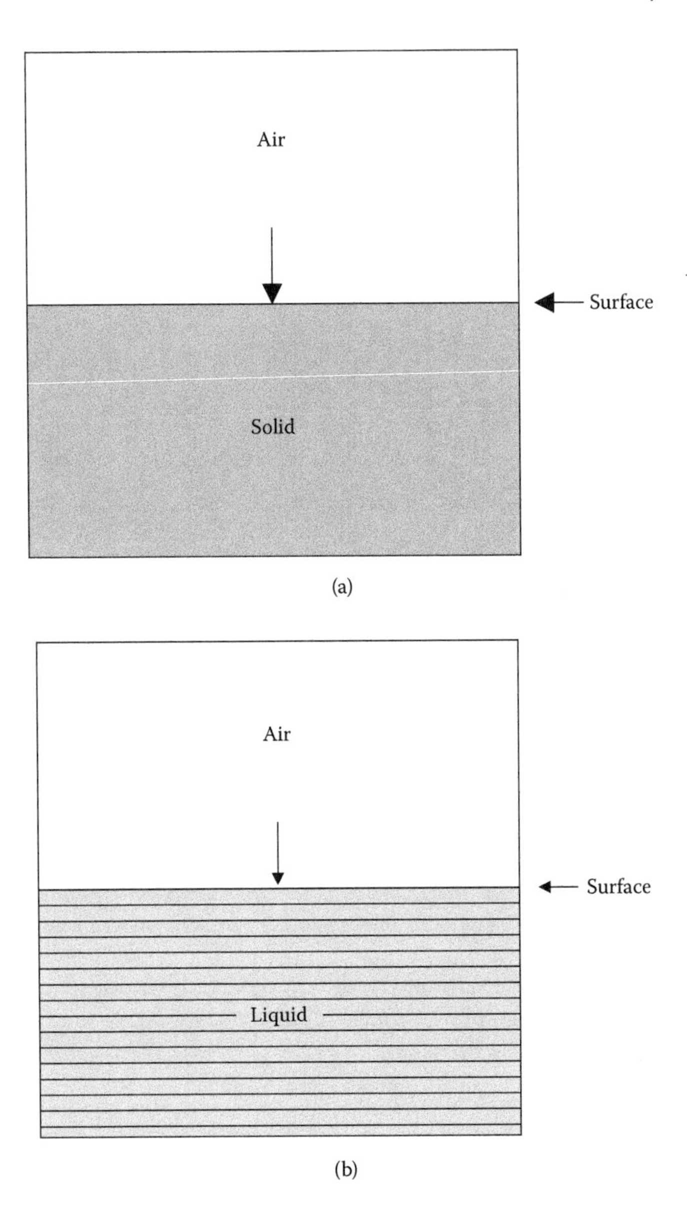

FIGURE 1.11 Different interfaces: (a) solid–gas (air), (b) liquid–gas (air), (c) solid–liquid, (d) solid 1–solid 2. (*Continued*)

system the fluid is uniform from the bulk phase to the surface. However, the *orientation* of the surface molecules will be different from those molecules in the bulk phase in all systems. For instance, in the case of water, the orientation of molecules inside the bulk phase will be different from those at the interface. The hydrogen bonding will orient the oxygen atom toward the interface. The question one may ask, then, is how sharply does the density change from that of being fluid to that of gas

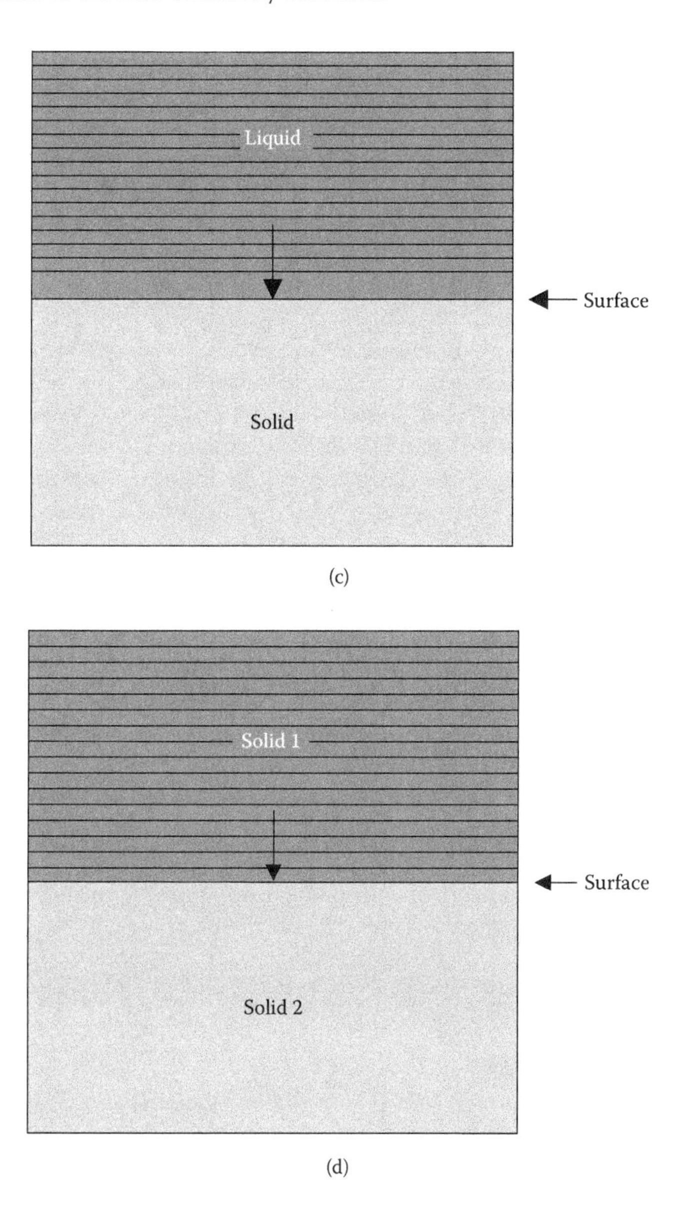

FIGURE 1.11 (*Continued*) Different interfaces: (a) solid–gas (air), (b) liquid–gas (air), (c) solid–liquid, (d) solid 1–solid 2.

(a change by a factor of 1000). Is this transition region a monolayer deep or many layers deep? Many studies have been reported where this subject has been investigated. *Gibbs adsorption theory* (Defay et al., 1966; Chattoraj and Birdi, 1984; Birdi, 1989, 1999, 2002, 2009, 2010a) considers the surface of liquids to be monolayer. The surface tension of water decreases appreciably with the addition of very small quantities of *soaps and detergents*. Gibbs adsorption theory relates the change in surface tension to the change in soap concentration. The experiments that analyze the spread monolayers are also based on one molecular layer. The latter data conclusively indeed verifies the Gibbs assumption (as described later). Detergents and other (soaps, etc.) similar kinds of molecules are found to exhibit self-assembly characteristics. The subject related to *self-assembly monolayer* (SAM) structures will be treated extensively (Birdi, 1999). However, no procedure that can provide information by a direct measurement exists. This subject will be described later herein. The composition of the surface of a solution with two components or more would require additional comments.

Colloids (the Greek word for "glue-like") are a wide variety of systems consisting of finely divided particles or macromolecules (such as, glue, gelatin, proteins) which are found in everyday life. Typical colloidal suspensions that are found in everyday life are provided in Table 1.1. Further, colloidal systems are widespread in their occurrence and have biological and technological significance. There are three types of colloidal systems (Adamson and Gast, 1997; Lyklema, 2000; Birdi, 2002, 2009; Dukhin and Goetz, 2002):

1. In simple colloids, clear distinction can be made between the disperse phase and the disperse medium, for example, simple emulsions of oil-in-water (O/W) or water-in-oil (W/O).
2. Multiple colloids involve the coexistence of three phases of which two are finely divided, for example, multiple emulsions (mayonnaise, milk) of water-in-oil-in-water (W/O/W) or oil-in-water-in-oil (O/W/O).
3. Network colloids have two phases forming an interpenetrating network, for example, polymer matrix.

The colloidal (in the form as solids or liquid drops) stability is determined by the free energy (surface free energy or the interfacial free energy) of the system. The

TABLE 1.1
Typical Colloidal Systems

Dispersed	Continuous	System Name
Liquid	Gas	Aerosol fog, spray
Gas	Liquid	Foam, thin films, froth, fire extinguisher foam
Liquid	Liquid	Emulsion (milk), mayonnaise, butter
Solid	Liquid	Sols, AgI, photography films, suspension wastewater, cement, oil recovery (shale oil), coal slurry
Corpuscles	Serum	Biocolloids (blood, blood coagulants)
Hydroxyapatite	Collagen	Bone, teeth
Liquid	Solid	Solid emulsion (toothpaste)
Solid	Gas	Solid aerosol (dust)
Gas	Solid	Solid foam (polystyrene), insulating foam
Solid	Solid	Solid suspension/solids in plastics

main parameter of interest is the large surface area exposed between the dispersed phase and the continuous phase. Since the colloid particles move about constantly, their dispersion energy is determined by the Brownian motion. The energy imparted by collisions with the surrounding molecules at temperature T = 300 K is 3/2 $k_B T$ = 3/2 1.38 10^{-23} 300 = 0.6 10^{-20} J (where k_B is the Boltzmann constant). This energy and the intermolecular forces would thus determine the colloidal stability.

In the case of colloid systems (particles or droplets), the kinetic energy transferred on collision will be thus $k_B\mathbf{T} = 10^{-20}$ J. However, at a given moment there is a high probability that a particle may have a larger or smaller energy. Further, the probability of total energy several times $k_B T$ (over 10 times $k_B T$) becomes very small. The instability will be observed if the ratio of the barrier height to $k_B T$ is around 1 to 2 units.

The idea that two species ($solid_1$–$solid_2$) should interact with one another, so that their mutual potential energy can be represented by some function of the distance between them, has been described in the literature. Furthermore, colloidal particles frequently adsorb (and even absorb) ions from their dispersing medium (such as in groundwater treatment and purification). Sorption that is much stronger than what would be expected from dispersion forces is called *chemisorption*, a process that is of both chemical and physical interest. For example, in a recent report it was mentioned that finely divided iron particles could lead to enhanced photosynthesis in oceans (resulting in the binding of large amounts of CO_2 from air). This could lead to control of the global warming effect. In fact, specific processes are being investigated that will lead to effective carbon capture (i.e., CO_2 capture) from such industries as coal (Krungleviclute et al., 2012).

As one knows from experience, oil and water do not mix (Figure 1.12), which suggests that these systems are dependent on the oil–water interface. The $liquid_1$–$liquid_2$ (oil–water) interface is found in many systems, most important is the world of *emulsions*.

The trick in using emulsions is based on the fact that one can apply both water and oil (the latter is insoluble in water) simultaneously. Further, one can then include other molecules that may be soluble in either phase (water or oil). This obviously leads to the common observation where we find thousands of applications of emulsions. It is very important to mention here that nature actually uses this trick in most of the major biological fluids. The most striking example is milk. The emulsion chemistry of milk is one of the most complex, and has still not been well investigated. Paint consists of polymer molecules dispersed in the water phase. After application, water evaporates leaving behind a glossy layer of paint.

In fact, the state of mixing oil and water is an important example of interfacial behavior at $liquid_1$–$liquid_2$. Emulsions of oil–water systems are useful in many aspects of daily life: milk, foods, paint, oil recovery, pharmaceutical, and cosmetics.

When olive oil is mixed with water, upon shaking, one will get the following:

- About 1 mm diameter oil drops are formed.
- After a few minutes the oil drops merge together and two layers (oil and water) are again formed.

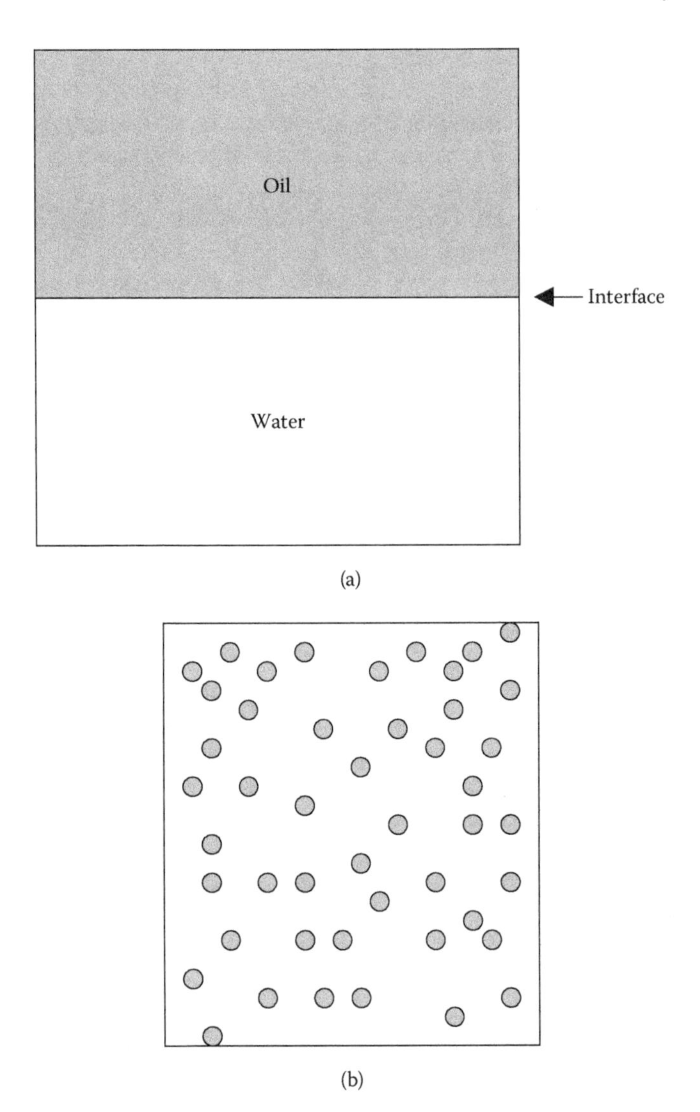

FIGURE 1.12 (a) Oil–water phases, (b) emulsion (oil drops mixed with water).

However, if suitable substances are added, which changes the surface forces, the olive oil drops that are formed can be very small (micrometer range). The latter leads to a stable emulsion.

The emulsion stability is basically stable depending on the size of the oil drops, in addition to other factors, dispersed in the water phase.

Low stability = large oil drops
Long stability = small drops
Very long stability = microsize drops

In addition, these considerations are important with regard to the different systems as follows: paints, cements, adhesives, photographic products, water purification, sewage disposal, emulsions, chromatography, oil recovery, paper and print industry, microelectronics, soap and detergents, catalysts, and biology (cell, virus). In some oil–surfactant–water–diverse components, liquid crystal (LC) phases (lyotropic LC) are observed. These lyotropic LC are indeed the basic building blocks in many applications of emulsions in technology. LC structures can be compared with a layer cake where each layer is molecular thick. It is thus seen that surface science pertains to investigations that take place at two different phases.

2 Capillarity and Surface Forces in Liquids (Curved Surfaces)

2.1 INTRODUCTION

It is known that liquids and solids have many many different physical characteristics. In this chapter some essential surface properties of liquids are described. One feature, which is well known, is that liquids take the shape of the container that surrounds or contains it. The question that arises then, is, what occurs when the liquid surface is *curved*, in comparison to a flat surface (Figure 2.1).

It may not seem that the curvature should impart different properties to a liquid (or solid), but surface chemistry research has added a very detailed analyses of these different structures (Goodrich et al., 1981). In fact, the extensive surface science research is mainly based upon the curvature effect, in addition to other properties. In the following systems it be seen that liquid surfaces are involved:

(a) The most common behavior is bubble and foam formation.
(b) Another phenomenon is that when a glass capillary tube is dipped in water, the fluid rises to a given height. It is observed that the narrower the tube, the higher the water rises.
(c) The role of liquids and liquid surfaces is important in many everyday natural processes (for example, oceans, lakes, rivers, raindrops). In everyday life, the most important liquid one uses is water. Oceans cover more than 75% of the earth's surface.

Further, the degree of curvature is found to impart different characteristics to liquid surfaces.

Another observation of great importance arises from the fact that the oceans cover some 75% of the surface of earth. Obviously, in this case the reactions on the surface of oceans will have much consequence for life on earth. For example, the carbon dioxide as found in air is distributed in various systems, such as in air, plants, oceans (CO_2 is soluble in water), and carbon capture process. Thus, CO_2 is in equilibrium in these different systems, which will have consequences on pollution control. In other words, it is not enough to monitor CO_2 in air alone; one must have the

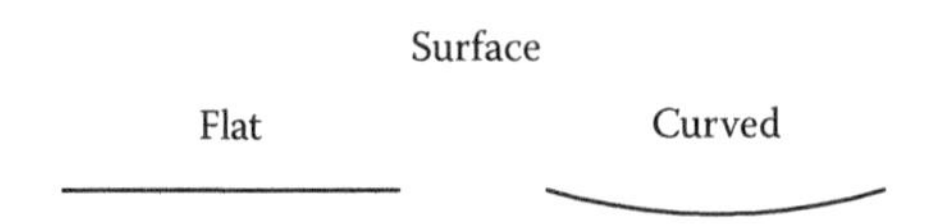

FIGURE 2.1 Flat or curved surfaces of a liquid.

knowledge of CO_2 concentrations in the other two states. Accordingly, one will thus need the study of surface tension and its effect on the surface phenomenon in these different systems. This means that the structures of molecules in the bulk phase in comparison to those at the surface need to be considered. In fact, capillary forces play a very important role in a variety of everyday life processes.

The state of molecules in different phases can be described as follows:

1. Molecules inside the bulk phase—Molecules are surrounded *symmetrically* in all directions.
2. Molecules at the surface—Molecules are interacting with molecules in the bulk phase (same as under 1), but toward the gas phase the interaction is weaker due to larger distances between molecules (Chapter 1, Figure 1.1).

The molecular forces that are present between the *surface molecules* are different from the molecules in the bulk phase or the gas phase. Accordingly, these forces are called *surface forces*. The surface forces make the liquid surface behave like a stretched elastic membrane in that it tends to contract. One cannot see this phenomenon directly, but it is observed through indirect experimental observations (both qualitatively and quantitatively). The latter arises from the observation that when one empties a beaker with a liquid, the liquid breaks up into spherical drops. This indicates that drops are being created under some forces that must be present at the surface of the newly formed interface. These surface forces become even more important when a liquid is in contact with a solid (such as groundwater, oil reservoir). The flow of liquid (e.g., water or oil) through small pores in the underground is mainly governed by the *capillary forces*. One can also accept that in order to recover oil from large depths (10 km) the underground needs appreciable pressure if the pores are very small. It has been found that capillary forces play a dominant role in many systems, which will be described later. Thus, the interaction between a liquid and any solid (with pores) will form curved surfaces, which being different from a planar fluid surface gives rise to the capillary forces. Another essential aspect is the mechanical surface tension present at *curved surfaces*. The curved liquid surfaces, such as in drops, or small capillaries, exhibit very special properties that differ from flat liquid surfaces. In this chapter the basics of surface forces will be described, and examples will be given where a system is dependent on these forces. This is essential since these principles are the building blocks for the understanding of this subject.

These surface forces interact both at $liquid_1$–$liquid_2$ (such as oil–water) and liquid–solid (such as cement–water) interfaces.

2.2 ORIGIN OF SURFACE FORCES IN LIQUIDS

The gas phase is converted to a liquid phase when the appropriate pressure and temperature is present (which means that the molecules become so close that a new phase, liquid, is formed). If the temperature is lowered, then the solid phase is formed. Liquid water freezes at 0°C to form solid ice (at 1 atmosphere). This kind of transition can be explained by analyzing the molecular interactions in liquids that are responsible for their physicochemical properties (such as *boiling point, melting point, heat of vaporization, surface tension*). Since molecular structure is the basic parameter involved in these transitions, the former characteristics need to be analyzed. These ideas are the basis for the *quantitative structure–activity relationship* (QSAR) (Birdi, 2002; Barnes, 2011). This requires one to understand the magnitudes of distance between molecules.

In the following, one can estimate the difference in molecular distances in liquid or gas as follows. In the case of water (for example), the following data is known (at room temperature and pressure):

Example: Water
Volume per mole liquid water = V_{liquid} = 18 ml/mole
Volume per mole water in gas state (at STP) (V_{gas}) = 22 liter/mole
Ratio $V_{gas}/V_{liquid} \cong 1000$

The approximate ratio of distance between molecules in a gas phase:liquid phase, will be $10 \cong (1000)^{1/3}$ (from simple geometrical considerations of volume [proportional to length^3] and length).

This shows that at the surface the density of water changes (Figure 2.2) from 1 ml/gram to 1000 ml/gram. In other words, the surface chemistry is related to those molecules that are situated in this transition region (Figure 2.3). Experiments have clearly shown that this transition region is of molecular dimension. The same is true for all liquids, with only minor differences.

In other words, the density of water changes 1000 times as the surface crosses from the liquid phase to the gas phase (air) (Figure 2.3). This means that in the gas phase, each molecule occupies 1000 times more volume than in the liquid phase. It means molecules move large distances before interacting with another molecule. This large change means that the surface molecules must be under a different environment than in the liquid phase or the gas phase. The distance between gas molecules is approximately 10 times larger than in a liquid. Hence, the forces (*all forces increase when distances between molecules decrease*) between gas molecules are much weaker than in the case of the liquid phase. All interaction

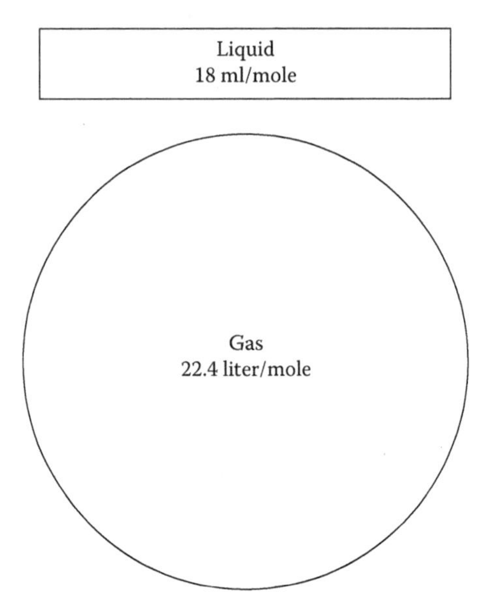

FIGURE 2.2 The volume of liquid water (18 ml/mole) and water in the vapor phase (as gas) (22.4 liter/mole).

forces between molecules (solid phase, liquid phase, gas phase) are related to the distance between molecules. The molecules at the surface thus have near-neighbor molecules much farther away than the molecules found inside the bulk of the liquid. It is also obvious that the surface of a liquid is a very busy place, since molecules are both evaporating into the air, and some are also returning into

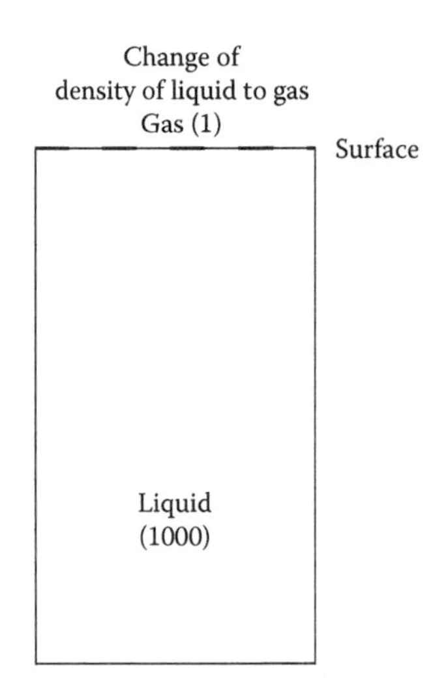

FIGURE 2.3 The change in density of a fluid (water) near the surface. (Distances between molecules is 10 times larger in the gas phase than in the liquid phase.)

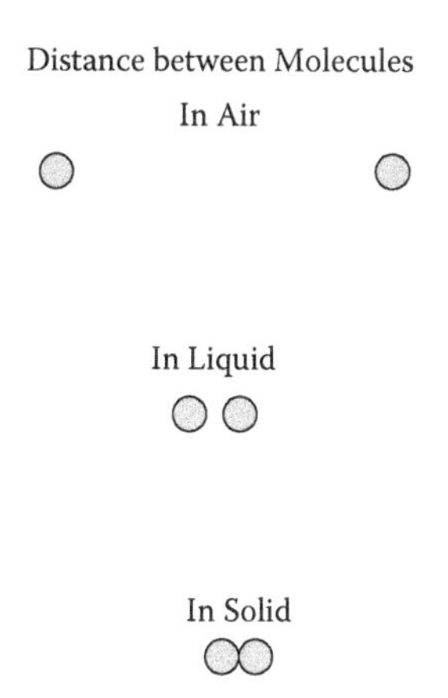

FIGURE 2.4 The distance between molecules in air, in liquids, in solids.

the liquid phase. This is a snapshot at the molecular level, so it cannot be seen by the naked eye.

The forces between molecules (or atoms) is determined by the distance between molecules. In the gas phase distances between molecules is roughly 10 times larger than in liquid or solid phase (Figure 2.4). In other words, the interaction energy between molecules in the gas phase is much weaker than in the liquid (or solid) phase.

In the case of water, it is the cohesive forces that maintain water, for example, in the liquid state at room temperature and pressure. It is useful as an example to compare cohesive forces in two different molecules, such as H_2O and H_2S. At room temperature and pressure, H_2O is liquid while H_2S is a gas. This means that H_2O molecules interact with different forces that are stronger and thus form a liquid phase. On the other hand, H_2S molecules exhibit much lower interactions and thus are in a gas phase at room temperature and pressure. In other words, the hydrogen bonds (between H and O) in water are stronger than hydrogen–sulfur bonds.

The molecules interact with each other, and the energy will be dependent on the geometrical packing (thus the magnitude of distances between molecules) in any given phase. The state of *surface energy* has also been described by the following classic example (Chattoraj and Birdi, 1984; Birdi, 1989; Adamson and Gast, 1997; Birdi, 1997, 2002, 2010a). Consider the area of a liquid film that is stretched in a wire frame by an increment dA, whereby the surface energy changes by (γ dA) (Figure 2.5). Under this process, the opposing force is f. From these data on dimensions we find:

Surface tension of a liquid = γ
Change in area = dA = l dx

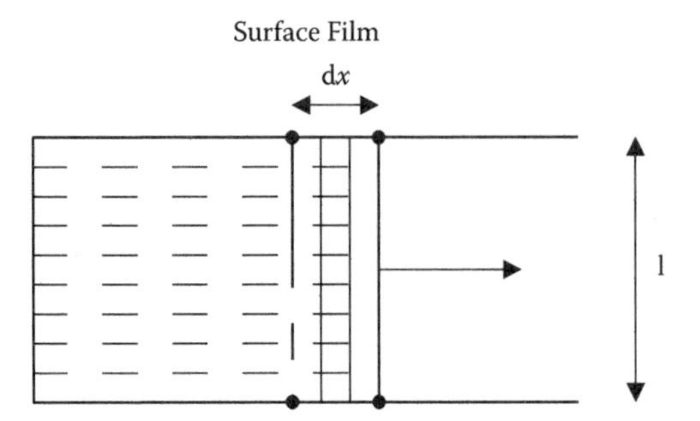

FIGURE 2.5 The surface film of a liquid. (See text for details.)

Change in x direction = dx

$$f\,dx = \gamma\,dA \quad (2.1)$$

or

$$\gamma = f\,(dx/dA) = f/2\,l \quad (2.2)$$

where dx is the change in displacement and l is the length of the thin film. The quantity γ represents the force per unit length of the surface (mN/m = dyne/cm), and this force is defined as surface tension or interfacial tension. *Surface tension*, γ, is the differential change of free energy with a change of surface area at a constant temperature, pressure, and composition.

Another example to describe surface energy may be considered. Let us imagine that a liquid fills a container that is the shape of a funnel. If the liquid in the funnel is moved upward then there will be an increase in *surface area*. This requires that some molecules from bulk phase have to move into the surface area and create extra surface A_s. The work required to do so will be (force times area) $\gamma\,A_S$. This is defined as reversible work (at a constant temperature and pressure), thus providing the increase in free energy of the system:

$$dG = -\,\gamma\,A_S \quad (2.3)$$

Thus, the tension per unit length in a single surface, or surface tension, γ, is numerically equal to the surface energy per unit area. Then G_s, the surface free energy per unit area:

$$G_s = \gamma = (d\,G/d\,A) \quad (2.4)$$

Under reversible conditions, the heat (q) associated with it gives the surface entropy, S_S:

$$dq = T\,dS_S \quad (2.5)$$

Combining these equations we find that:

$$d\gamma/dT = -S_s \quad (2.6)$$

Further

$$H_S = G_S + T_S \tag{2.7}$$

One can also write for surface energy, E_S:

$$E_S = G_S + T\,S_S \tag{2.8}$$

These relations give:

$$E_S = \gamma - T\,(d\gamma/dT) \tag{2.9}$$

The quantity E_S has been found to provide more useful information on surface phenomena than any of the other quantities (Birdi, 2009).

Here, we find that the term S_s is the surface entropy per square centimeter of surface. This shows that to change the surface area of a liquid (or solid, as described later), a *surface energy* exists (γ, surface tension), which needs to be considered. The quantity γ means that to create 1 m^2 (= 10^{20} $Å^2$) of new surface of water, one will need to use 72 mJ energy. To transfer a molecule of water from the bulk phase (where it is surrounded by about 10 near neighbors by about 7 k_BT) (k_BT = 4.12 10^{-21} J) to the surface, one needs to break about half of these hydrogen bonds (i.e., 7/2 k_B T = 3.5 k_BT). The free energy of transfer of one molecule of water (with area of 12 $Å^2$) will be about 10^{-20} J (or about 3 k_BT). This is a reasonable quantity under these simple geometrical assumptions.

In the case of solid systems, similar consideration is needed if one increases the surface area of a solid (for example by crushing) (Figure 2.6). The *surface tension of the solid* needs to be measured and analyzed. It has been found that the energy needed to crush a solid is related to the surface forces (i.e., solid surface tension). More examples as given later will provide real-world situations where γ of both liquids and solids is needed to describe the physics of the surfaces.

2.3 CAPILLARY FORCES: LAPLACE EQUATION, AND LIQUID CURVATURE AND PRESSURE (MECHANICAL DEFINITION)

Liquids show specific characteristics due the fact that molecules in liquids are able to move inside a container, whereas a solid cannot exhibit this property. It has been found that this property of liquid gives some specific properties, such as curved surfaces in narrow tubings (Goodrich et al., 1981). It is interesting to consider aspects of the wettability. Surely everyone has noticed that water tends to rise near the walls of a glass container. This happens because the molecules of this liquid have a strong tendency to adhere to the glass. Liquids that wet the walls make concave surfaces (e.g., water/glass), those that do not wet them, make convex surfaces (e.g., mercury/glass). Inside tubes with internal diameters smaller than 2 mm, called capillary tubes, a wettable liquid forms a concave meniscus in its upper surface and tends to

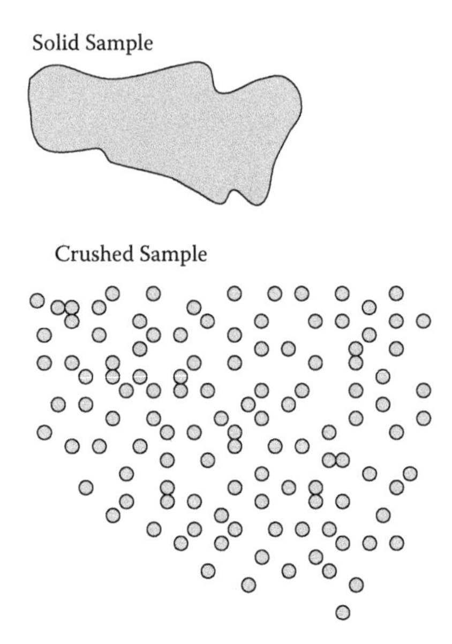

FIGURE 2.6 The solid sample and crushed powder (such as talcum and cement).

go up along the tube. On the contrary, a nonwettable liquid forms a convex meniscus and its level tends to go down. The amount of liquid attracted by the capillary rises until the forces that attract it balance the weight of the fluid column. The rising or the lowering of the level of the liquids into thin tubes is named *capillarity* (*capillary force*). Also the capillarity is driven by the forces of cohesion and adhesion, as we have already mentioned. Capillary forces play a very important role in a wide range of everyday life. These examples will be delineated wherever such systems are encountered in this book.

It should be noted that a liquid inside a large beaker has almost a flat surface. However, the same liquid inside a fine tubing will be curved (Figure 2.7). This

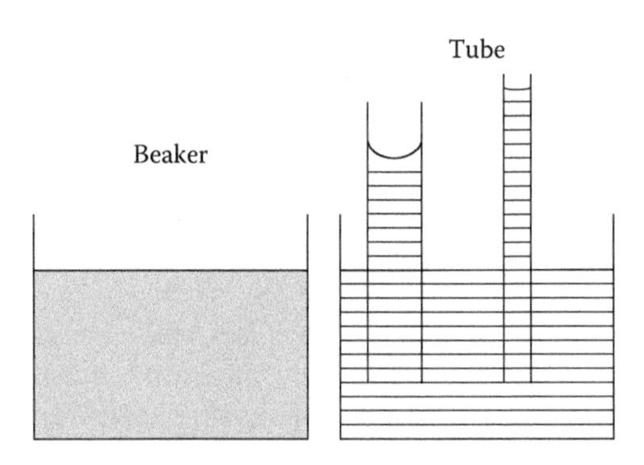

FIGURE 2.7 The surface of water inside a large beaker and in a narrow tubing. The rise of water in the tubing is due to the capillary force.

behavior is very important in everyday life. The physical nature of this phenomena is the subject of this section. In other words, the curved liquid surface gives rise to some very characteristic properties, which are found in everyday life.

The surface tension, γ, and the mechanical equilibrium at interfaces has been described in the literature in detail (Chattoraj and Birdi, 1984; Birdi, 1989, 2002, 2009, 2010a; Adamson and Gast, 1997). The surface has been considered to be a hypothetical stretched *membrane*, this membrane being termed as the surface tension. In a real system undergoing an infinitesimal process, it can be written:

$$dW = p\, dV + p'\, dV' - \gamma\, dA \tag{2.10}$$

where dW is the work done by the systems when a change in volume, dV and dV′, occurs; p and p' are pressures in the two phases α and ß, respectively, at equilibrium; and dA is the change in interfacial area. The sign of the interfacial work is designated negative by convention (Chattoraj and Birdi, 1984).

The fundamental property of liquid surfaces is that they tend to contract to the smallest possible area. This is not observed in the case of solids. This property is observed in the spherical form of small drops of liquid, in the tension exerted by soap films as they tend to become less extended, and in many other properties of liquid surfaces. In the absence of gravity effects, these curved surfaces are described by the Laplace equation, which relates the mechanical forces as (Chattoraj and Birdi, 1984; Adamson and Gast, 1997; Birdi, 1997):

$$p - p' = \gamma\,(1/r_1 + 1/r_2) \tag{2.11}$$

$$= 2\,(\gamma/r) \tag{2.12}$$

where r_1 and r_1 are the radii of curvature (in the case of an ellipse), while r is the radius of curvature for a spherical shaped interface. It is a geometric fact that surfaces for which Equation (2.11) hold are surfaces of minimum area. These equations thus give:

$$dW = p\, d(V + V') - \gamma\, dA \tag{2.13}$$

$$= p\, dV^t - \gamma\, dA \tag{2.14}$$

where $p = p'$ for plane surface, and V^t is the total volume of the system.

It has been found that a pressure difference across the curved interfaces of liquids (such as drops or bubbles) exists. For example, if a tube is dipped into water (or any fluid) and a suitable amount of pressure is applied, then a bubble is formed (Figure 2.8). This means that the pressure inside the bubble is greater than the atmosphere pressure. It thus becomes apparent that curved liquid surfaces induce effects that need special physicochemical analyses in comparison to flat liquid surfaces. It should be noted that in this system, a mechanical force has induced a change on the surface of a liquid. This phenomena is also called *capillary forces*. Then one may ask does this also require similar consideration in the case of solids. The answer is yes and will be discussed in detail later. For example, in order to remove a liquid that is inside a porous media such as a sponge, a force equivalent to these capillary forces would be needed. Man has been fascinated with bubbles for many centuries. As can

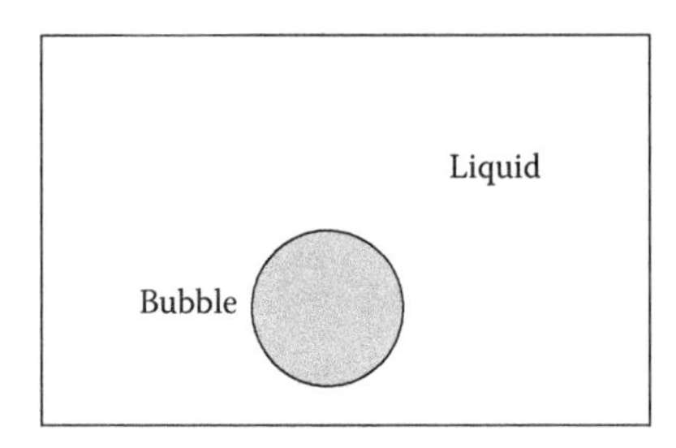

FIGURE 2.8 An air bubble in a liquid. The pressure inside the bubble is greater than the pressure outside the bubble.

be seen in Figure 2.4, the bubble is produced by applying a suitable pressure, ΔP, to obtain a bubble of radius R, where the surface tension of the liquid is γ.

It is useful to consider the phenomena where the bubble is expanded by applying pressure, P_{inside}. In this process, there are two processes to consider: (1) the surface area of the bubble will increase by dA; and (2) the volume will increase by dV. In other words, there are two opposing actions: the expansion of volume and the increase of surface area. The work done can be expressed in terms of that done against the forces of surface tension and that done by increasing the volume. At equilibrium, the following condition between these two kinds of work exists:

$$\gamma\, dA = (P_g - P_{liquid})\, dV \tag{2.15}$$

where $dA = 8\pi R\, dR$ ($A = 4\pi R^2$), and $dV = 4\pi R^2\, dR$ ($V = 4/3\pi R^3$). Combining these relations gives the following:

$$\gamma 8\pi R\, dR = \Delta P 4\pi R^2\, dR \tag{2.16}$$

and

$$\Delta P = 2\,\gamma/R \tag{2.17}$$

where $\Delta P = (P_g - P_{liquid})$. Since the free energy of the system at equilibrium is constant, $dG = 0$, then these two changes in system are equal. If the same consideration is applied to the soap bubble, then the expression for ΔP bubble will be:

$$\Delta P_{bubble} = p_{inside} - p_{outside} = 4\,\gamma/R \tag{2.18}$$

because now two surfaces exist and a factor of 2 is needed to consider this state.

The pressure applied gives rise to work on the system, and the creation of the bubble gives rise to creation of surface area increase in the fluid. The Laplace equation relates the pressure difference across any curved fluid surface to the curvature, 1/radius and its surface tension, γ. In those cases where nonspherical curvatures are present, one obtains the more universal equation:

$$\Delta P = \gamma\,(1/R_1 + 1/R_2) \tag{2.19}$$

It is also seen that in the case of spherical bubbles, since $R_1 = R_2$, this equation becomes identical to Equation 2.18. It is thus seen that in the case of a liquid drop in

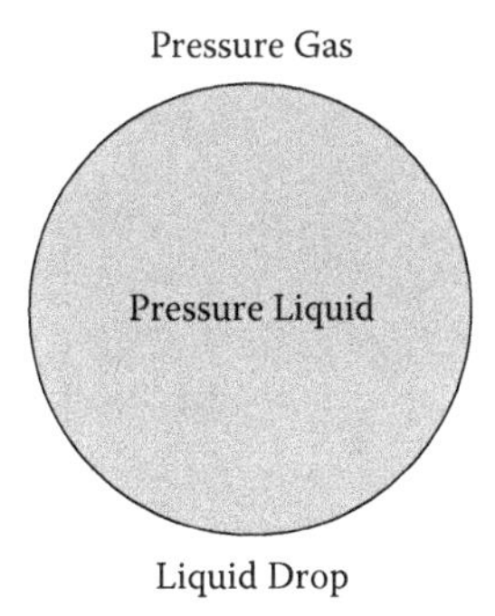

FIGURE 2.9 The liquid drop and ΔP (= $P_{LIQUID} - P_{GAS}$) (at a curved liquid–gas interface).

air (or gas phase), the Laplace pressure would be the difference between the pressure inside the drop, p_L, and the gas pressure, p_G (Figure 2.9):

$$p_L - p_G = \Delta P \tag{2.20}$$

$$= 2\ \gamma/\text{Radius} \tag{2.21}$$

In the case of a water drop with a radius of 2μ, there will be ΔP of magnitude:

$$\Delta P = 2\ (72\ \text{mN/m})/2\ 10^{-6}\ \text{m} = 72000\ \text{Nm}^{-2} = 0.72\ \text{atm} \tag{2.22}$$

It is known that pressure changes the vapor pressure of a liquid. Thus, ΔP will affect the vapor pressure and lead to many consequences in different systems. In fact, the capillary (Laplace) pressure determines many industrial and biological systems. The lung alveoli are dependent on the radii during the inhale–exhale process, and the change in surface tension of the fluid lining the lung alveoli. In fact, many lung diseases are related to the lack of surface pressure and capillary pressure balance. Blood flow through arteries of different diameters throughout the body is another system where Laplace pressure is of much interest for analytical methods (such as heart function and control). The Laplace equation is useful for analysis in a variety of systems, for example:

1. Bubbles or drops (raindrops or combustion engines, sprays, fog)
2. Blood cells (flow of blood cells through arteries)
3. Oil or groundwater movement in rocks
4. Lung vesicles

In the following, some typical examples are given where the Laplace equation is found to play an important role. Another important conclusion is that ΔP is larger inside a small bubble than in a larger bubble with the same γ. This means that when two bubbles meet, the smaller bubble will enter the larger bubble to create a new bubble (Figure 2.10). This phenomenon can have important consequences in various systems (such as emulsion stability, lung alveoli, oil recovery, bubble characteristics [such as in champagne and beer]). The same is observed when two liquid drops contact each other, the smaller will merge into the larger drop.

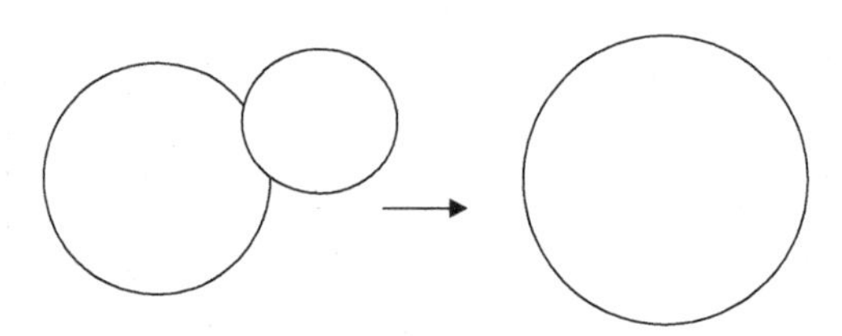

FIGURE 2.10 Coalescence of two bubbles with different radii.

In Figure 2.11 is a system that initially shows that two bubbles of different curvature are connected through a regulator (which can be closed or open). After the tap is opened, the smaller bubble shrinks while the larger bubble (with lower ΔP) increases in size until equilibrium is reached (when the curvature of the two bubbles become equal in magnitude). This kind of equilibrium is the basis of lung alveoli where fluids (containing lipid surfactants) balance out the expanding–contracting cycle (Birdi, 1989).

It has been observed that a system with varying size bubbles collapse faster than bubbles (or liquid drops) that are of exactly the same size. In other words, in systems with bubbles of similar diameter there will be a slower coalescence than in systems with varying sized bubbles. Another major consequence is observed in the

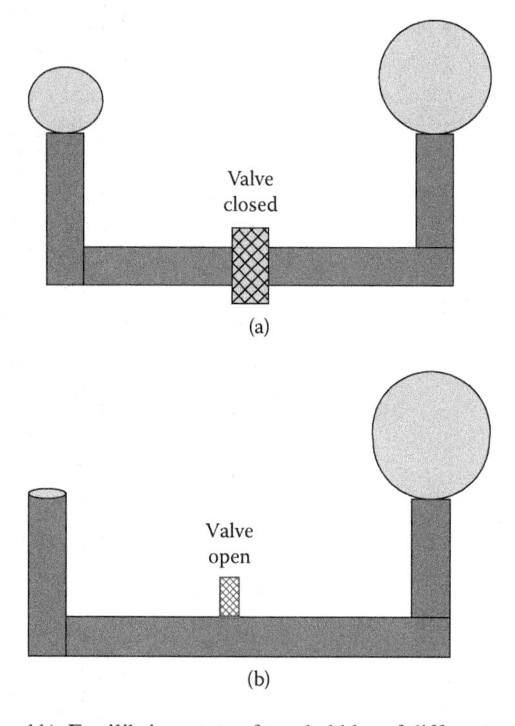

FIGURE 2.11 (a and b): Equilibrium state of two bubbles of different radii. (See text.)

oil recovery phenomena (see later). Oil production takes place (in general) by applying gas or water injection. When gas or water injection is applied, where there are small pores the pressure needed will be higher than in the large pore zone. Thus the gas or water will bypass the small pore zone and leave the oil behind (at present more than 30% to 50% of oil in place is not recovered under normal production methods). This is obviously a great challenge to the surface and colloid chemists in the future. Enhanced oil recovery (EOR) investigations will be described later in this book (Birdi, 2002).

In industry one can monitor the magnitude of surface tension with the help of bubble pressure. Air bubbles are pumped through a capillary into the solution. The pressure measured is calibrated to known γ solutions, thus by using a suitable computer one can estimate γ values very accurately. Commercial apparatus are available to monitor surface tension. Liquid curved surfaces are found in many different everyday life phenomena. It is interesting to consider the consequences of Laplace pressure in different curved surfaces. One important example is that when a small drop comes in contact with a larger drop, the former will merge into the latter. Another aspect is that vapor pressure over a curved liquid surface, p_{cur}, will be larger, than on a flat surface, p_{flat}. A relation between pressure over curved and flat liquid surface was derived (Kelvin equation):

$$\ln (p_{cur}/p_{flat}) = (v_L/R\ T)\ (2\ \gamma/R_{cur}) \tag{2.23}$$

where p_{cur} and p_{flat} are the vapor pressures over curved and flat surfaces, respectively. R_{cur} is the radius of curvature and v_L is the molar volume. The Kelvin equation thus suggests that if liquid is present in a porous material, such as cement, then the difference in vapor pressure exists between two pores of different radii. Similar consequence of vapor pressure exists when two solid crystals of different size are concerned. The smaller sized crystal will exhibit higher vapor press and will also result in a faster solubility rate.

Raindrop formation from clouds: The transition from water vapor in clouds to raindrops is not as straightforward a process as it might seem. The formation of large liquid raindrops requires that a certain number of water molecules in the clouds (as gas phase) have formed a nuclei (which is the formation of first liquid drop). This is observed in some cases where there are many clouds but not enough rain, which is needed for irrigation (this happens if nuclei are absent). The nuclei or embryo will grow and the Kelvin relation will be the determining factor. Artificial rain has been attempted by using fine particles which lead to nuclei formation and assisting in raindrop formation.

2.4 CAPILLARY RISE OR FALL OF LIQUIDS (CAPILLARY FORCES)

The surface tension of liquids becomes evident when one observes the following experiment. Capillary forces are the reason that liquids behave differently when a narrow capillary tube is dipped into a liquid (Figure 2.12). Liquids in narrow tubes

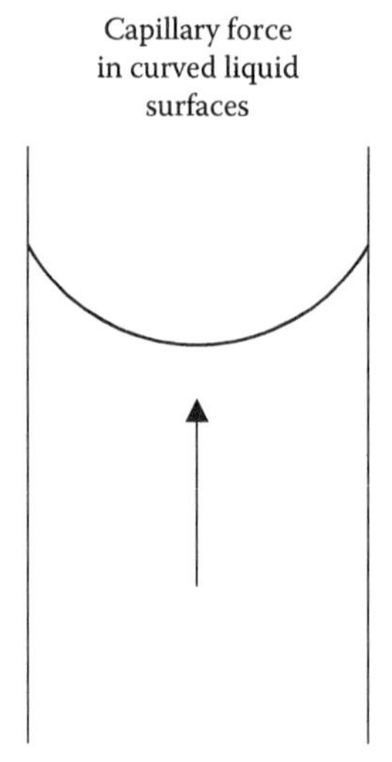

FIGURE 2.12 Curved liquid surface and capillary forces.

are found in many different technical and biological systems. The range of these applications is from the blood flow in the veins to the oil recovery in the reservoir. Fabric properties are also governed by capillary forces (that is wetting). A sponge absorbs water or other fluids where the capillary forces push the fluid into the many pores. This is also called a *wicking* process (as in candlewicks).

The curvature in a system where a narrow capillary circular tube is dipped into a liquid exhibits properties that are not observed in a large beaker. The liquid is found to rise in the capillary, when the fluid wets the capillary (such as water and glass, or water and metal). The *curvature* of the liquid inside the capillary will lead to pressure difference between this state and the relatively flat surface outside the capillary (Figure 2.13).

The rise or fall of a liquid in a capillary (arising from the capillary force) is dependent upon the wetting characteristics. Inside the capillary, the liquid (with surface tension, γ), attains an equilibrium of capillary forces. However, if the fluid is nonwetting (such as mercury [Hg] in glass) then one finds that the fluid *falls*. This arises from the fact that Hg does not wet the tube. Capillary forces arise from the difference in attraction of the liquid molecules to each other and the attraction of the liquid molecule to those of the capillary tube. The fluid rises inside the narrow tube to a height, h, until the surface tension forces balance the weight of the fluid. This equilibrium gives the following relation:

$$\gamma 2\pi R = \text{Surface tension force} \quad (2.24)$$

$$= \rho_L g_g h\pi R^2 = \text{Fluid weight} \quad (2.25)$$

where γ is the surface tension of the liquid and R is the radius of curvature. In the case of narrow capillary tubes, less than 0.5 mm, the curvature can be safely set

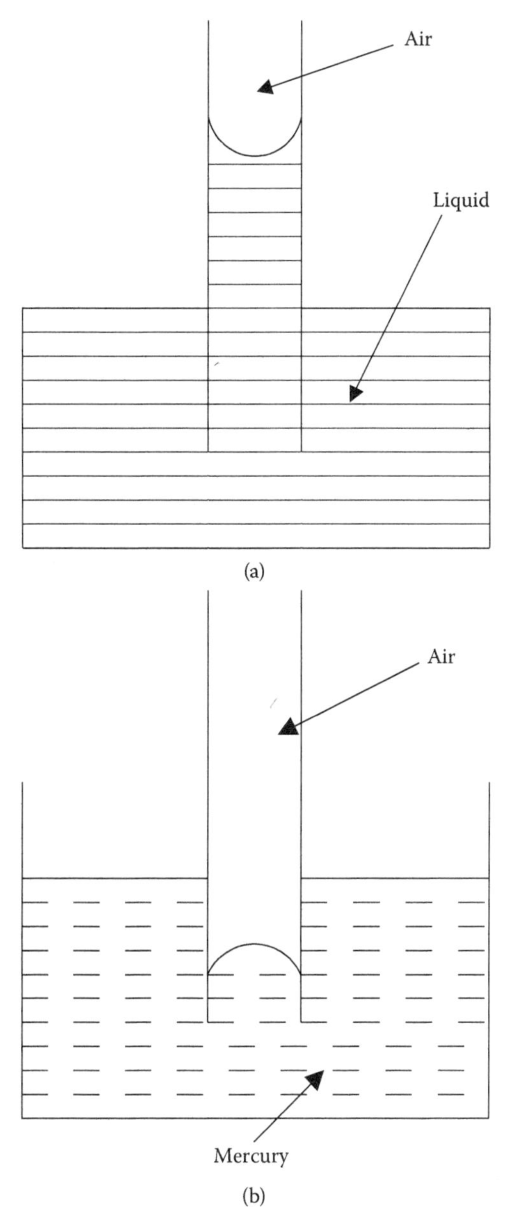

FIGURE 2.13 The rise of a liquid in a glass capillary (a) of water, and the (b) fall of Hg in a capillary.

equal the radius of the capillary tubing. The fluid will rise inside the tube to compensate for surface tension force, thus at equilibrium we get:

$$\gamma = {}^1/_2 R \rho_L g h \tag{2.26}$$

where ρ_L is the density of the fluid, g is the acceleration of gravity, and h is the rise in the tube.

Example 2.1

Liquid = water (25° C)
Water ($\gamma = 72.8$ mN/m; $\rho_L = 1000$ kg/m^3; $g = 9.8$ m/s^2) will rise to a height of

Magnitude of capillary rise in different size capillary:

$$h = \gamma / {}^1/_2 R \rho_L g$$
$$= 72.8/{}^1/_2 R\ (1000)\ 9.8$$

The magnitudes for h for different radii of tubings are:

0.015 mm in a capillary of radius 1m
1.5 mm in a capillary of radius 1 cm
14 cm in a capillary of radius 0.1 mm

One can measure the magnitude of h very accurately, and thus allows one to measure the value of γ.

It has been found that these assumptions are only precise when the capillary tubing is rather small. In the case of larger-sized capillaries, correction tables are found in the literature (Adamson and Gast, 1997). This phenomenon is important for plants with very high stems where water is needed for growth. However, in some particular cases where the contact angle, θ, is not zero, a correction is needed, and the equation will become:

$$\gamma = 2\ R\rho_L gh\ (1/\cos(\theta)) \tag{2.27}$$

When the liquid wets the capillary wall, the magnitude of θ is 0, and cos(0) = 1. In the case of Hg, the contact angle is 180°, since it is a nonwetting fluid (see Figure 2.13). Since cos(180) = –1, then the sign of h in the equation will be negative. This means that Hg will show a drop in height in glass tubing. Hence, the rise or fall of a liquid in a tubing will be governed by the sign of cos(θ).

Thus, capillary forces play an important factor in all systems where liquids are present in a porous environment. Similar results can also be derived by using the Laplace equation (Equation 2.21) (1/radius = 1/R):

$$\Delta P = 2\ \gamma / R \tag{2.28}$$

The liquid rises to a height h, and the system achieves equilibrium, and the following relation is found:

$$2\ \gamma / R = h g_g \rho_L \tag{2.29}$$

This can be rewritten:

$$\gamma = 2\ R\ \rho_L\ g_g\ h \tag{2.30}$$

Thus, it can be seen that the various surface forces are responsible for the capillary rise. The lower the surface tension, the lower the height of column in the capillary. The magnitude of γ is determined from the measured value h for a fluid with known ρ_L. The magnitude of h can be measured directly by using a suitable device (for example, a photograph image).

Further, it is known that in the real world, capillaries or pores are not always circular in shape. In fact, one considers that in oil reservoirs the pores are more like triangular or square shaped than circular. In this case, one can measure the rise in other kinds of shaped capillaries, such as rectangular or triangular (Birdi et al., 1988; Birdi, 1997, 2002). These studies have much importance for oil recovery or water treatment systems. Especially in shale oil/gas technology, large amounts of water are used to transport chemicals through pores where curved liquid surfaces are present. In any system where fluid flows through porous material, it is expected that capillary forces will be one of the most dominant factors. Further, it is known that the vegetable world is dependent on capillary pressure (and osmotic pressure) to bring water up to the higher parts of plants. In this way, some trees succeed in bringing this essential liquid (water) up to 120 meters above the ground.

2.5 SOAP BUBBLES: FORMATION AND STABILITY

Perhaps the phenomenon of bubble formation is the most common observation mankind has experienced since childhood. Bubbles are also commonly observed in many different instances:

- Beer and champagne
- Along the coasts of lakes and oceans
- Shampoo and detergent solutions

Extensive research has been carried out with regard to bubble formation (Boys, 1959; Lovett, 1994; Birdi, 1997; Taylor, 2011). It is well known that soap bubbles are extremely thin (a thousand times thinner than the diameter of a hair!) and unstable. In spite of the latter, under special conditions one can keep soap bubbles for long lengths of time, which thus allows one to study its physical properties (such as thickness, composition, conductivity, spectral reflection, etc.). In most cases, the thickness of a bubble is over hundreds of micrometers in the initial state. The film consists of a bilayer of detergent that contains the solution. The film thickness decreases with time due to the following reasons: (1) drainage of fluid away from the film and (2) due to evaporation.

Therefore, the stability and lifetime of such thin films will be dependent on these different characteristics. This is found from the fact as an air bubble is blown under the surface of a soap or detergent solution, the air bubble will rise up to the surface. It may remain at the surface, if the speed is slow, or it may escape into the air as a soap bubble. Experiments show that a soap bubble consists of a very thin liquid film with an iridescent surface. But as the fluid drains away and the thickness decreases, the latter approaches to the equivalent of barely two surfactant molecules plus a few molecules of water. It is worth noting that the limiting thickness is of the order of two or more surfactant molecules. This means that one can see with the naked eye the molecular size structures of thin liquid films (if curved). As the air bubble enters the surface region, the soap molecules along with water molecules are pushed up and as

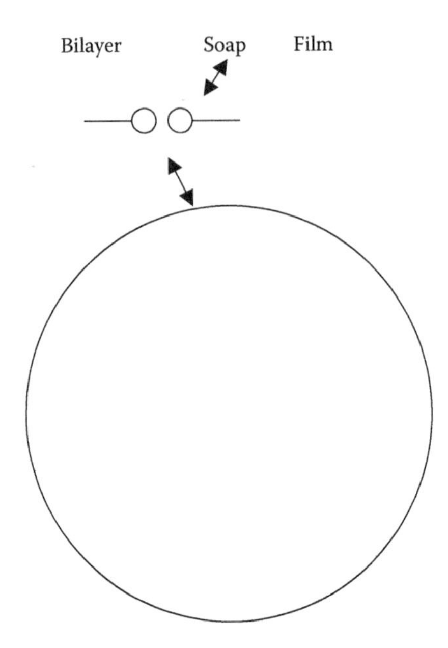

FIGURE 2.14 A bilayer soap film structure.

the bubble is detached, it leaves a thin-liquid film with the following characteristics (as found from various measurements):

- A bilayer of soap (approximately 200 Å thick) on the outer region contains the aqueous phase (Figure 2.14).
- The thickness of the initial soap bubble is some micrometers.
- The thickness decreases with time and one starts to observe rainbow colors, as the reflected light is of the same wavelength as the thickness of the bubble (few hundreds of angstroms). The thinnest liquid film consists mainly of the bilayer of surface active substance (such as soap = 50 Å) and some layers of water. The light interference and reflection studies show many aspects of these thin-liquid films.

The bilayer soap film may be depicted as arrays of the soap molecule (with a few layers of water) (— — -O: length of the soap molecule is about 15 Å):

Thick bubble film (micometer or more) (shows no colors):

```
— — -OWWWWWO— — -
— — -OWWWWWO— — -
— — -OWWWWWO— — -
```

Thin bubble film (shows colors):

```
— — -OWO— — -
— — -OWO— — -
— — -OWO— — -
```

The thickness of the bilayer is thus about 30 to 50 Å (almost twice the length of the soap molecule). The thickness is approximately the size of wavelength of light.

The iridescent colors of the soap bubble arise from the interference of reflected light waves. The reflected light from the outer surface and the inner layer give rise to this interference effect (Figure 2.15). The rainbow colors are observed as the bubble thickness decreases (and reaches magnitudes corresponding to that in the light waves) due to the evaporation of water.

Rainbow colors (violet, indigo, blue, green, yellow, orange, red) are observed when the thickness is of the order of wavelength of light. One may remember this sequence of colors as: VIBGYOR. The wavelength increases from violet to red color. One observes rainbow colors in all sorts of situations where interference of waves of light takes place. Another interesting observation is that in some natural phenomena, colors are produced by simply structures made similar to the bubble, that is, thickness (of lipid-like molecules) varies in the range of wavelength of light. This is, for instance, the case in the colors observed in peacock feathers (and many other colored objects in nature and otherwise).

Thicker films (thickness is mainly due to the water) reflect red light and one observes blue-green colors. Lesser thin films cancel out the yellow wavelength and blue color is observed. As the thickness approaches the wavelength of light, all

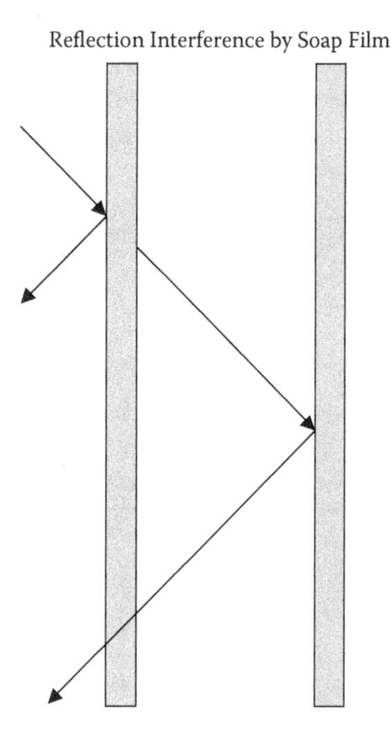

FIGURE 2.15 The reflection of light waves from the outer layer and inner layer. The interference is caused by the difference in the path of light traveling inside the thin film.

colors are canceled out and a black (or gray) film is observed. This corresponds to 25 nm (250 Å).

The transmitted light, I_{tr}, is related to the incident, I_{in}, and the reflected intensity, I_{re}:

$$I_{tr} = I_{in} - I_{re} \tag{2.31}$$

As seen from the analyses of the daylight, it consists of different wavelengths of colors (violet, indigo, blue, green, yellow, orange, red):

Red, 680 nm
Orange, 590 nm
Yellow, 580 nm
Green, 530 nm
Blue, 470 nm
Violet, 405 nm

Slightly thicker soap films (ca. 150 nm) sometimes look golden. In the thinning process the different colors get cut off. Thus if the blue color gets cut off, the film looks amber to magenta.

Soap bubble solutions: In general one may use a detergent solution (1 to 10 g/liter detergent concentration). These solutions are sufficient to blow bubbles that may or may not last for long times. However, to produce stable bubbles (which means slow evaporation rates, stability to vibration effects, etc.) then some additives have been used. A common recipe is given next:

- Detergent (dishwashing), 1 to 10 g/L
- Glycerin (reduces evaporation), less than 10%
- Water, rest

Another recipe has the following composition (Lovett, 1994): 100 g glycerin, 1.4 g triethanolamine, and 2 g oleic acid. Actually, glycerin can be added to ordinary household detergent solutions, which may also give stable bubbles.

The bubble film (which consists of surface active substance + water + salts) stability can be described as follows:

- Evaporation of water (making thin film and unstable)
- Flow of water away from the film
- Stability of the bubble film (to vibration or other mechanical action)

It is thus obvious that the rate of water evaporation plays an important role. The evaporation can be reduced by containing the bubble in a closed bottle. One also finds that in such a closed system the bubbles remain stable for very long time. The drainage of water away from the film is dependent on the viscosity of the fluid. Therefore such additives as glycerin (or other thickening agents [polymers]) assist in the stability.

2.6 MEASUREMENT OF SURFACE TENSION OF LIQUIDS

In the literature, a variety of methods used to measure the magnitude of surface tension of liquids can be found. This arises from the fact that one needs a specific method for each situation, which one may use in the measurement of γ. For example, if the liquid is water (at room temperature) then the method will be different than if the system is molten metal (at a very high temperature, ca. 500°C or higher). In the oil reservoirs one finds oil at high temperatures (over 80°C) and pressures (over 200 atm). In many cases, specific instruments that allow for measuring the magnitude of surface tension under the given situation have been developed. Some of the common methods used will be described in this section.

2.6.1 Shape and Weight of Liquid Drops

The formation of liquid drops when flow occurs through thin tubes is a common daily phenomenon. In some cases, such as eyedrop application, the size of drop plays a significant role in the application and dosage of medicine (such as eyedrop solutions). The drop formed when liquid flows through a circular tube is shown in Figure 2.16.

In many processes (such as oil recovery, blood flow, underground water) one encounters liquid flow through thin (micrometer diameter) noncircular shaped tubes or pores. Studies that address these latter systems can be found in the literature. In other contexts, the liquid drop formation, for example, in an ink-jet nozzle: this

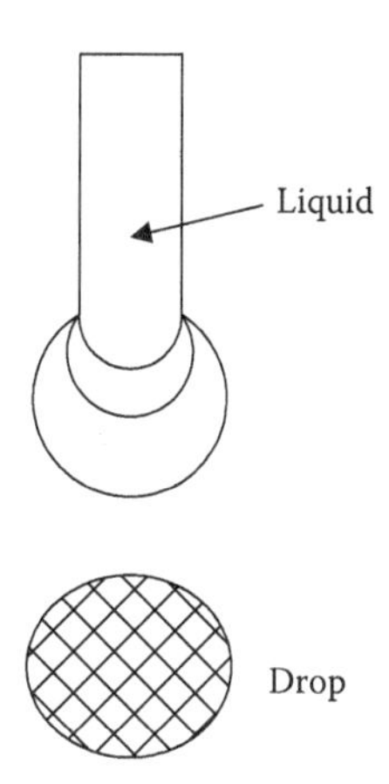

FIGURE 2.16 Liquid drop formation through a tube.

technique falls under a class of scientifically challenging technology. The ink jet printer demands are of such quality that this branch of drop-on-demand technology is much industrial research and technology. All combustion engines are controlled by the oil drop formation and evaporation characteristics. The important role of capillary forces is obvious in such systems. As the liquid drop grows larger it will at some stage break off the tube (due to gravity force being larger than the surface force holding it to the capillary) and will correspond to the maximum weight of the drop that can hang. The equilibrium state where the weight of the drop is exactly equal to the detachment surface energy is given as:

$$m_m\, g = 2\pi R\gamma \tag{2.32}$$

where m_m is the weight of the detached drop, and R is the radius of the tubing.

A simple method is to count the number of drops (for example 10 or more) and measure the weight (with a very high accuracy).

One may also use a more convenient method where a fluid is pumped and the drops are collected and weighed. Since in some systems (solutions) there may be kinetic effects one must be careful to keep the flow as slow as possible. This system is very useful in studying systems found in daily life phenomena: oil flow, blood cells, flow through arteries, and so on. In those cases where the volume of fluid available is limited, one may use this method with advantage. By decreasing the diameters of the tubing one can work with less than 1 μL fluids. This may be case by case of such systems as eye fluids.

One can determine the magnitude of γ from either the maximum weight or from the shape of the drop. The "detachment" method is based upon detaching a body from the surface of a liquid, which wets the body. It is necessary to overcome the same surface tension forces that operate when a drop is broken away. The liquid attached to the solid surface on detachment creates the following surfaces:

Initial stage—Liquid attached to solid
Final stage—Liquid separated from solid

In the process, from the initial stage to the final stage, the liquid molecules that were near the solid surface have moved away and are now near their own molecules. This requires energy, and the force required to make this happen is proportional to the surface area of contact and to the surface tension of the liquid. However, their advantage over the latter method consists in that it makes it possible to choose the most convenient form and size of the body (platinum rod, ring or plate) so as to enable the measurement to be carried out rapidly but without any detriment to its accuracy. The detachment method has found an application in the case of liquid whose surface tension changes with time.

In some cases, the amount of liquid available is very small, such as fluid from the eye. Under these conditions, one finds that following procedure is most suitable for the measurement of γ. The liquid drop forms as it flows through a tubing (Figure 2.17). At a stage just before it breaks off, the shape of the *pendant drop* has been used to estimate γ. The drop shape is photographed and from the diameters of the shape one can accurately determine γ. Actually, if one has only a drop of fluid, then one can measure its γ without loss of sample volume.

The parameters needed are as follows. A quantity pertaining to the ratio of two significant diameters are:

$$S_s = d_s/d_e \tag{2.33}$$

where d_e is the equatorial diameter, and d_s is the diameter at a distance d_e from the tip of the drop (Figure 2.17). The relation between γ and d_e and S_s is found as:

$$\gamma = \rho_L \, g \, d_e^{\,2}/H \tag{2.34}$$

where ρ_L is the density of the liquid and H is related to S_s, but the values of 1/H for varying S_s were obtained from experimental data. For example when $S_s = 0.3$, $1/H = 7.09837$, while when $S_s = 0.6$, then $1/H = 1.20399$. Accurate mathematical functions have been used to estimate 1/H for a given d_e value (Adamson and Gast, 1997; Birdi, 2002). The accuracy (0.1%) is satisfactory for most of the systems. Especially when experiments are carried out under extreme conditions (such as: high temperatures and pressures).

The pendant drop method is very useful under specific conditions:

1. Technically, only a drop (a few microliters) is required. For example, eye fluid can be studied since only a drop of a microliter is needed.
2. It can be used under very extreme conditions (very high temperatures or corrosive fluids).
3. Under very high pressure and temperatures. Oil reservoirs are found typically at 100°C and 300 atm pressure. Surface tension of such systems can be conveniently studied by using high pressure and temperature cells with optical clear windows (sapphire windows 1 cm thick, up to 2000 atm). For example, γ of inorganic salts at high temperatures (ca. 1000°C) can be measured by using this method. The variation of surface tension can be studied as a function of various parameters (temperature and pressure, additives [gas, etc.]).

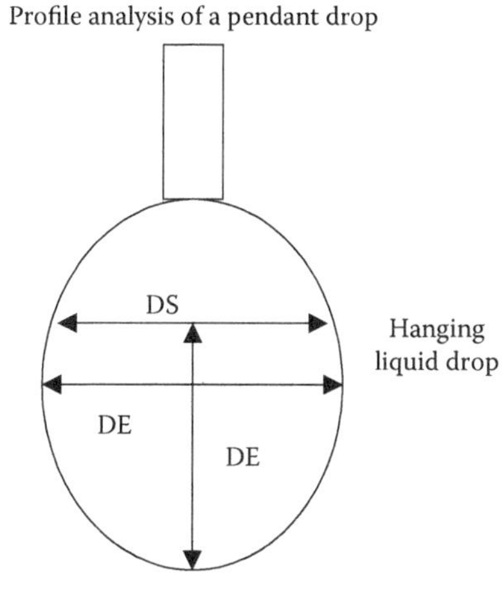

FIGURE 2.17 A pendant drop of liquid (shape analysis).

2.6.2 The Ring Method (Detachment)

In the classical methods used to measure surface tension of liquids we find that the detachment of a solid from a liquid surface provides very accurate results. A method that has been rather widely used involves the determination of the force to detach a ring or loop of wire from the surface of a liquid. This method is based on using a ring (platinum) and measuring the force when it is dipped in the liquid surface.

This method is one of the many detachment methods, of which the drop weight and the Wilhelmy slide method are also examples. This method was originally developed by du Nouy. As with all detachment methods, one supposes that within an accuracy of a few percents, the detachment force is given by the surface tension multiplied by the periphery of the surface (liquid surface) detached (from a solid surface of a tubing or ring or plate). This assumption is also acceptable for most experimental purposes. Thus, for a ring, as illustrated in Figure 2.18,

$$W_{total} = W_{ring} + 2\,(2\pi R_{ring})\,\gamma \tag{2.35}$$

$$= W_{ring} + 4\pi R_{ring}\,\gamma \tag{2.36}$$

where W_{total} is the total weight of the ring, W_{ring} is the weight of the ring in air, R_{ring} is the radius of the ring. The circumference is $2\pi R_{ring}$, and a factor of 2 is because of the two sides of contact.

This relation assumes that the contact between the fluid and the ring is geometrically simple. It has also been found that this relation is fairly correct (better than 1%) for most working situations. However, it was observed that Equation 2.36 needed a correction factor, in much the same way as was done for the drop weight method. Here, however, there is one additional variable so that the correction factor f now depends on two dimensionless ratios.

Experimentally, this method is capable of good precision. A so-called chainomatic balance has been used to determine the maximum pull, but a popular simplified version of the tensiometer, as it is sometimes called, makes use of a torsion wire and is quite compact. Among experimental details to mention are that the dry weight of the ring, which is usually constructed of platinum, is to be used, the ring should be kept horizontal (a departure of 1° was found to introduce an error of 0.5%, whereas one

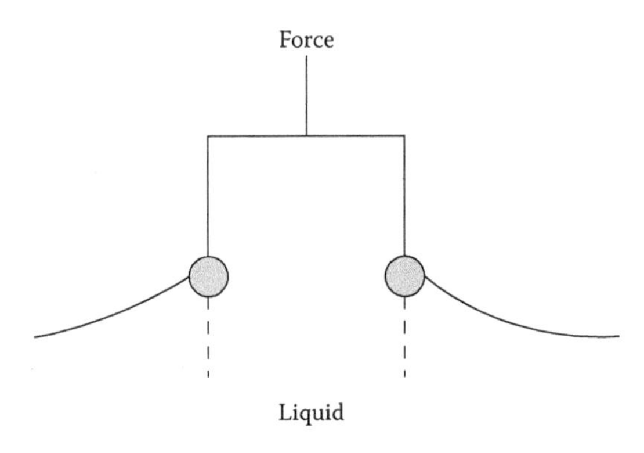

FIGURE 2.18 The ring method. (See text for details.)

of 2.1° introduced an error of 1.6%), and care must be taken to avoid any disturbance of the surface as the critical point of detachment is approached (Adamson and Gast, 1997). The ring is usually flamed before use to remove surface contaminants such as grease, and it is desirable to use a container for the liquid that can be overflowed so as to ensure the presence of a clean liquid surface. A zero or near-zero contact angle is necessary; otherwise results will be low. This has been the case with surfactant solutions where adsorption on the ring changed its wetting characteristics, and where liquid–liquid interfacial tensions were measured. In such cases, a Teflon or polyethylene ring may be used.

2.6.3 Plate (Wilhelmy) Method

The methods discussed thus far have required more or less tabular solutions, or else correction factors to the respective "ideal" equations. Further, if continuous measurements need to be made, then it is not easy to use some of these methods (such as the capillary rise or bubble method). The most useful method for measuring surface tension is the well-known Wilhelmy plate method. If a smooth and flat plate shaped metal is dipped in a liquid, the surface tension forces will give rise to a tangential force (Figure 2.19). This is because a new contact phase is created between the plate and the liquid.

The total weight measured, W_{total}, would be:

$$W_{total} = \text{Weight of the plate} + \gamma\,(\text{perimeter}) - \text{Updrift} \tag{2.37}$$

The perimeter of a plate is the sum of twice the length plus breadth. The surface force will act along the perimeter of the plate, that is, Length (L_p) + Width (W_p). The plate is often very thin (less than 0.1 mm) and made of platinum, but even plates made of glass, quartz, mica, and filter paper can be used. The forces acting on the plate consist of the gravity and surface tension downward, and buoyancy due to displaced water upward. For a rectangular plate of dimensions L_p and W_p and of material density ρ, immersed to a depth h_p in a liquid of density ρ_L, the net downward

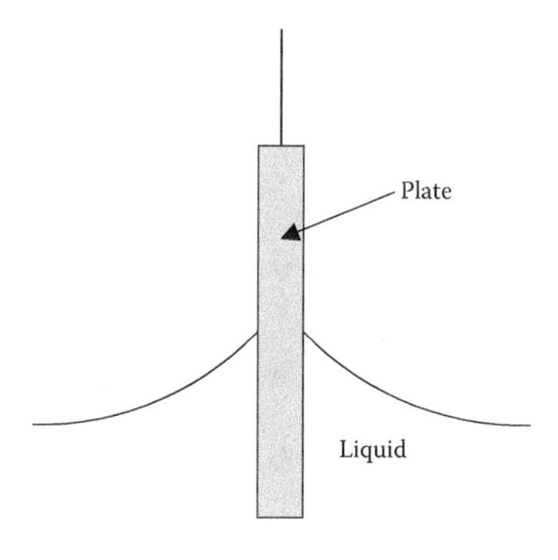

FIGURE 2.19 Wilhelmy plate in a liquid. (Plate with dimensions: length = L_p, width = W_p.)

force, F, is given by the following equation (i.e., Weight of plate + Surface force ($\gamma \times$ Perimeter of the plate – Upward drift)):

$$F = \rho_p\, g\, (L_p W_p t_p) + 2\, \gamma\, (t_p + W_p)(\cos(\theta)) - \rho_L\, g\, (t_p W_p h_p) \qquad (2.38)$$

where γ is the liquid surface tension, θ is the contact angle of the liquid on the solid plate, and g is the gravitational constant. If the plate used is very thin (i.e., $t_p << W_p$) and the updrift is negligible (i.e., h_p is almost zero), then γ is found from:

$$\gamma = (F)/(2\, W_p) \qquad (2.39)$$

The sensitivity of γ by using these procedures was found to be very high ($\pm$ 0.001 dyne/cm (mN/m)) (Birdi, 2009). The change in surface tension (surface pressure = Π) is then determined by measuring the change in F for a stationary plate between a clean surface and the same surface with a monolayer present. If the plate is completely wetted by the liquid (i.e., $\cos(\theta) = \cos(0) = 1$) the surface pressure is then obtained from the following equation:

$$\Pi = -[\Delta\, F/2(t_p + W_p)] = -\Delta\, F/2\, W_p, \quad \text{if } W_p >> t_p \qquad (2.40)$$

In general, by using very thin plates with thickness 0.1 to 0.002 mm, one can measure surface tension with very high sensitivity. The apparatus is calibrated using pure liquids, such as water and ethanol. The buoyancy correction is made very small (and negligible) by using a very thin plate and dipping the plate as little as possible. The wetting of water on platinum plate is achieved by using commercially available platinum plates, which are roughened to increase wettability. The latter property gives rise to almost complete wetting, that is, $\theta = 0$. In this way, the force is determined by measuring the changes in the mass of the plate, which is directly coupled to a sensitive electrobalance or some other suitable device (such as a pressure transducer). It should be noted that in all systems where an object is in contact with a liquid, there will exist a force toward the liquid. In other words, in everyday life real examples, one will observe a variety of such systems.

2.7 TYPICAL SURFACE TENSION DATA OF LIQUIDS

At this stage, it is important to consider how the magnitude of surface tension of different molecules changes with respect to the molecular structure. Extensive studies are found that attempt to correlate surface tension to other physicochemical properties of liquids, such as boiling point and heat of evaporation. This concept has been extensively analyzed in literature (Birdi, 2010a). In research and other applications where surface tension of liquids plays an important role, it is necessary to be able to predict the magnitude of γ of different kinds of molecules (Table 2.1). A brief analyses of these data is given in the following.

These data require an explaination of the differences in surface tension data and molecular structure. The range of γ is found to vary from ca. 20 to over 1000 mN/m. The surface tension of Hg is high (425 mN/m), for example as compared to that of water, because it is a liquid metal with a very high boiling point. The latter indicates

TABLE 2.1
Surface Tension Values of Some Common Liquids

Liquid	Surface Tension
1,2-Dichloro-ethane	33.3
1,2,3-Tribromo propane	45.4
1,3,5-Trimethylbenzene (Mesitylene)	28.8
1,4-Dioxane	33.0
1,5-Pentanediol	43.3
1-Chlorobutane	23.1
1-Decanol	28.5
1-nitro propane	29.4
1-Octanol	27.6
Acetone (2-Propanone)	25.2
Aniline	43.4
2-Aminoethanol	48.9
Anthranilic acid ethylester	39.3
Anthranilic acid methylester	43.7
Benzene	28.9
Benzylalcohol	39.0
Benzylbenzoate (BNBZ)	45.9
Bromobenzene	36.5
Bromoform	41.5
Butyronitrile	28.1
Carbon disulfide	32.3
Quinoline	43.1
Chloro benzene	33.6
Chloroform	27.5
Cyclohexane	24.9
Cyclohexanol	34.4
Cyclopentanol	32.7
p-Cymene	28.1
Decalin	31.5
Dichloromethane	26.5
Diiodomethane (DI)	50.8
1,3-Diiodopropane	46.5
Diethylene glycol	44.8
Dipropylene glycol	33.9
Dipropylene glycol monomethylether	28.4
Dodecyl benzene	30.7
Ethanol	22.1
Ethylbenzene	29.2
Ethylbromide	24.2
Ethylene glycol	47.7

(Continued)

TABLE 2.1 (Continued)
Surface Tension Values of Some Common Liquids

Liquid	Surface Tension
Formamide	58.2
Fumaric acid diethylester	31.4
Furfural (2-furaldehyde)	41.9
Glycerol	64.0
Ethylene glycol monoethyl ether (ethyl cellosolve)	28.6
Hexachlorobutadiene	36.0
Iodobenzene	39.7
Isoamylchloride	23.5
Isobutylchloride	21.9
Isopropanol	23.0
Isopropylbenzene	28.2
Isovaleronitrile	26.0
m-Nitrotoluene	41.4
Mercury	425.4
Methanol	22.7
Methyl ethyl ketone (MEK)	24.6
Methyl naphthalene	38.6
N,N-dimethyl acetamide (DMA)	36.7
N,N-dimethyl formamide (DMF)	37.1
n-Decane	23.8
n-Dodecane	25.3
n-Heptane	20.1
n-Hexadecane	27.4
n-Hexane	18.4
n-Octane	21.6
n-Tetradecane	26.5
n-Undecane	24.6
n-Butylbenzene	29.2
n-Propylbenzene	28.9
Nitroethane	31.9
Nitrobenzene	43.9
Nitromethane	36.8
o-Nitrotoluene	41.5
Perfluoroheptane	12.8
Perfluorohexane	11.9
Perfluorooctane	14.0
Phenylisothiocyanate	41.5
Phthalic acid diethylester	37.0
Polyethylen glycol 200 (PEG)	43.5
Polydimethyl siloxane	19.0
Propanol	23.7

TABLE 2.1 (Continued)
Surface Tension Values of Some Common Liquids

Liquid	Surface Tension
Pyridine	38.0
3-Pyridylcarbinol	47.6
Pyrrol	36.6
sym-Tetrabromoethane	49.7
tert-Butylchloride	19.6
sym-Tetrachloromethane	26.9
Tetrahydrofuran (THF)	26.4
Thiodiglycol	54.0
Toluene	28.4
Water	72.8
o-Xylene	30.1
m-Xylene	28.9
a-Bromonaphthalene (BN)	44.4
a-Chloronaphthalene	41.8
Metals (liquid state, high temperature)	Greater than 1000

Notes: Surface tension (20°C; mN/m [dyne/cm]).

that it needs much energy to break the bonds between Hg atoms to evaporate. Similarly, γ of NaCl as a liquid (at high temperatures) is also very high. The same is found for metals in a liquid state. The other liquids can be considered under each type, which should help to understand the relationship between the structure of a molecule and its surface tension.

Alkanes—The magnitude of γ increases by 1.52 mN/m per two $-CH_2$, when alkyl chain length increases from 10 to 12 (n-Decane, 23.83; n-Dodecane, 25.35).

n-Heptane, 20.14
n-Hexadecane, 27.47
n-Hexane, 18.43
n-Octane, 21.62

Alcohols—The magnitude of γ changes by 23.7 – 22.1 = 1.6 mN/m per -CH2-group. This is based upon the γ data of ethanol (22.1 mN/m) and propanol (23.7 mN/m).
These observations indicate the molecular correlation between bulk forces and surface forces (tension)(γ) for homologous series of substances.

2.7.1 Effect of Temperature and Pressure on the Surface Tension of Liquids

All natural processes are dependent on the temperature and pressure effects on any system under consideration. For example, oil reservoirs are generally found under high temperatures (about 100°C) and pressure (over 200 atm). Actually, mankind is aware of the great variations of both temperature (sun) and pressure (earthquakes, storms, and winds) with which the natural phenomena are surrounding the earth. Even the surface of the earth itself comprises a temperature variation of –50°C to 50°C. On the other hand, inside the center mantle of earth increases in temperature and pressure as one goes from its surface to the center of earth (about 5000 km). In fact, the scientific research comprises from temperatures ranging from minus 273°C (absolute temperature) to over 2000°C (such as the inside of the earth or the surface of sun). The surface tension is related to the internal forces in the liquid (surface), and one must thus expect it to bear relationship to the internal energy. Further, it has been found that surface tension always decreases with increasing temperatures. Surface tension, γ, is a quantity that can be measured accurately and applied in the analyses of all kinds of surface phenomena. If a new surface is created, then in the case of a liquid, molecules from the bulk phase must move to the surface. The work required to create extra surface area, dA, will be given as:

$$dG = \gamma\, dA \tag{2.41}$$

The surface free energy, G_S, per unit area is given as:

$$G_S = \gamma = (dG/dA)_{T,P} \tag{2.42}$$

Hence, the other thermodynamic surface quantities will be:

$$\text{Surface entropy: } S_S = -(d\, G_S/dT)_P \tag{2.43}$$

$$= -(d\, \gamma/dT) \tag{2.44}$$

We can thus derive for surface enthalpy, H_S:

$$H_S = G_S + TS_S \tag{2.45}$$

All natural processes are dependent on the effect of temperature and pressure. For instance, oil reservoirs are found under high temperatures (about 80°C) and pressure (around 100 to 400 atm depending on the depth). Thus, one must investigate such systems under these parameters. This is related to the fundamental equation for the free energy, G, and to the enthalpy, H, and entropy, S, of the system:

$$G = H - TS \tag{2.46}$$

Thus, it is important that in all practical analyses one should be aware of the effects of temperature and pressure. The molecular forces that stabilize liquids will be expected to decrease as the temperature increases. Experiments also show that in all cases, surface tension decreases with increasing temperatures.

Surface entropy of liquids is given by $(-d\, \gamma/dT)$. This means that the entropy is positive at higher temperatures (because the magnitude always decreases with temperature for all liquids). The rate of decrease of surface tension with temperature is different for different liquids, which supports the aforementioned description of liquids.

For example, the surface tension of water data is given as:

at 5°C, $\gamma = 75$ mN/m
at 25°C, $\gamma = 72$ mN/m
at 90°C, $\gamma = 60$ mN/m

Extensive γ data for water was fitted to the following equation (Birdi, 2002):

$$\gamma = 75.69 - 0.1413\ t_C - 0.0002985\ t_C^2 \tag{2.47}$$

where t_C is in Celsius. This equation gives the value of γ at 0°C as 75.69 mN/m. The value of γ at 50°C is found to be $(75.69 - 0.1413 \times 50 - 0.0002985 \times 50 \times 50)$ = 67.88 mN/m, and at 25°C it is 71.97 mN/m.

In the literature, such relationships for other liquids, which allows one to calculate the magnitude of γ at different temperatures can be found. This kind of application is important in the oil industry.

Further, these data show that γ of water decreases with temperatures from 25°C to 60°C (72 – 60)/(90 – 25) = 0.19 mN/m °C. This is what should expected from physicochemical theory. In all systems, as the temperature increases the energy between molecules gets weaker and thus the surface energy, that is, surface tension, will decrease. The difference in the surface entropy gives information on the structures of different liquids. It is also observed that the effect of temperature will be lower for liquids with a higher boiling point (such as Hg) than for low boiling liquids (such as n-hexane). Actually, a correlation exists between the surface tension and the heat of vaporization (or boiling point). In fact, many systems even show big differences when comparing winter or summer months (such as raindrops, sea waves, foaming in natural environments). Different thermodynamic relations have been derived that can be used to estimate the surface tension at different temperatures. Especially, straight chain alkanes have been extensively analyzed. The data show that a simple correlation exists between surface tension, temperature, and n_C. This allows one to estimate the value of any temperature of a given alkane. This observation has many aspects in the applied industry. It allows one to quickly estimate the magnitude of surface tension of an alkane at the required temperature. Further, this has a fairly good quantitative analysis about how surface tension will change for a given alkane under a given experimental condition. A detailed analyses is given in Appendix 2A.

2.7.2 Different Physical Aspects of Liquid Surfaces

Liquid surfaces exhibit many different kinds of physical properties, which need to be mentioned. In the following, only the essential aspects are provided and the reader is encouraged to look at more advanced literature for further information, if needed.

2.7.3 Heat of Liquid Surface Formation and Evaporation

All matter is stabilized by forces interacting between molecules. Therefore, as one moves inside a liquid phase toward the surface, the consequences of the changing interaction forces need to be considered, which are related to the surrounding structure. In order to understand which forces stabilize liquid structures, a relation between the surface tension of a liquid to the latent heat of evaporation has been suggested. This is a reasonable argument considering the geometrical packing of molecules. It was argued a century ago (Stefan, 1886) that when a molecule is brought to the surface of a liquid from the interior, the work done in overcoming the attractive force near the surface should be related to the work expended when it escapes into the scarce vapor phase (Adamson and Gast, 1997; Birdi, 1997, 2002). It was suggested that the first quantity should be approximately half of the second. According to the Laplace theory of capillarity, the attractive force acts only over a small distance equal to the radius of the sphere (see Figure 1.5 in Chapter 1), and in the interior the molecule is attracted equally in all directions and experiences no resultant force. On the surface, it experiences a force due to the liquid in the hemisphere, and half the total molecular attraction is overcome in bringing it there from the interior.

A very useful molecular model suggested by Stefan (1886) was that the energy necessary to bring a molecule from the bulk phase to the surface of a liquid should be half the energy necessary to bring it entirely into the gas phase. As known from geometrical considerations, a sphere can be surrounded by 6 molecules (in two-dimension) (Figure 2.20) and 12 molecules (in three-dimension) of the same size.

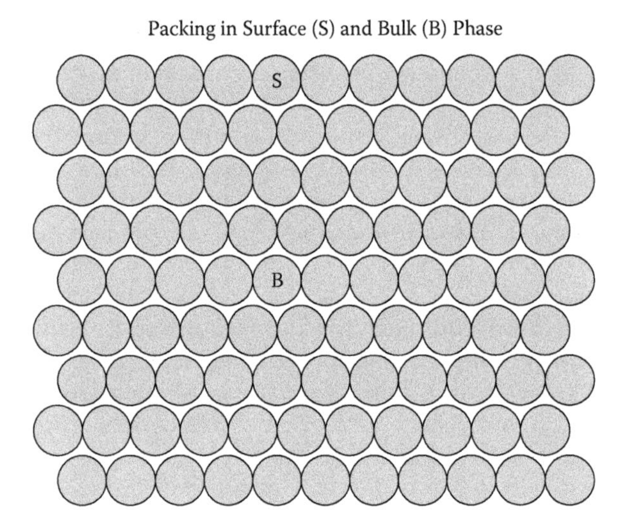

FIGURE 2.20 Packing of molecules (in two-dimension) in the bulk phase (six neighbors) and near the surface (three neighbors). (See text for details.)

TABLE 2.2
Ratio of Enthalpy of Surface Formation, h_s, and the Enthalpies of Vaporization, h_{vap}

Molecules (Liquid)	h_s/h_{vap}
Hg	0.64
N_2	0.51
O_2	0.5
CCl_4	0.45
C_6H_6	0.44
Diethyl ether	0.42
$Cl\text{-}C_6H_5$	0.42
Methyl formate	0.40
Ethyl acetate	0.4
Acetic acid	0.34
H_2O	0.28
C_2H_5OH	0.19
CH_3OH	0.16

This corresponds with the most densely packed top (surface) monomolecular layer half filled and the next layer completely filled to a very dilute gas phase (the distance between gas molecules is approximately 10 times greater than in liquids or solids). This indicates that intermolecular forces in liquids would be weaker than in solids by a few orders of magnitude, as is also found experimentally. The ratio of the enthalpy of surface formation to the enthalpy of vaporization, h_s: h_{vap}, for various substances is given in Table 2.2, substances with nearly spherical shaped molecules have ratios near 1/2, while substances with a polar group on one end give a much smaller ratio. This difference indicates that the latter molecules are oriented with the nonpolar end toward the gas phase and the polar end toward the liquid. In other words, molecules with dipoles would be expected to be oriented perpendicularly at the gas/liquid interfaces. It is important to note that such simple considerations add appreciable knowledge to systems such as surface phenomena.

A simple analysis was proposed that was based on purely spherical geometrical packing of molecules (most of which are certainly not spherical) and ideal situations. Hence, any deviation from Stefan's law is an indication that the surface molecules are oriented differently than in the bulk phase. This observation is useful in order to understand surface phenomena.

As an example, one may proceed with this theory and estimate the surface tension of a liquid with data on its heat of evaporation. The number of near neighbors of a surface molecule will be about half (6 = 12/2) than those in the bulk phase (12 neighbors). It is now possible to estimate the ratio of the attractive energies in the bulk and in the surface, per molecule. We have following data for a liquid such as CCl_4:

$$\text{Molar energy of vaporization} = \Delta U_{vap} \tag{2.48}$$

$$= \Delta h_{vap} - RT$$

$$= 34000\ Jmol^{-1} - 8.315\ J\ K^{-1}\ mol^{-1}\ (298\ K)$$

$$= 31522\ J\ mol^{-1} \tag{2.49}$$

$$\text{Energy change per molecule} = 31522\ Jmol^{-1}/6.023\ 10^{23}\ mol^{-1}$$

$$= 5.23\ 10^{-20}\ J \tag{2.50}$$

If we assume that about half of energy is gained when a molecule is transferred to the surface, then we get:

$$\text{Energy per molecule at surface} = 5.23\ 10^{-20}\ (2)\ J = 2.6\ 10^{-20}\ J \tag{2.51}$$

The molecules at the surface occupy a certain value of area, which can be estimated only roughly as follows.

$$\text{Density of } CCl_4 = 1.59\ g\ cm^{-3}$$

$$\text{Molar mass} = 12 + 4(35.5) = 154\ g\ mol^{-1}$$

$$\text{Volume per mol} = 154/1.59 = 97\ cm^3\ mol^{-1}$$

$$\text{Volume per molecule} = 97\ 10^{-6}\ m^3\ mol^{-1}/6.023\ 10^{23}\ mol^{-1} = 1.6\ 10^{-28}\ m^3$$

$$\text{The radius of a sphere (volume} = 4/3\ \Pi\ R^3\text{) with this magnitude of volume}$$

$$= [1.6\ 10^{-28}/(4/3\ \Pi)]^{1/3} = 3.5\ 10^{-10}\ m$$

$$\text{Area per molecule} = \Pi\ R^2 = \Pi\ (3.5\ 10^{-10})^2 = 38\ 10^{-20}\ m^2$$

$$\text{Surface tension (calculated) for } CCl_4 = 2.6\ 10^{-20}\ J/38\ 10^{-20}\ m^2$$

$$= 0.068\ J\ m^{-2} = 68\ mNm^{-1}$$

The measured value of γ of CCl_4 is 27 mN/m (Table 2.1). The large difference can be ascribed to the assumption that Stefan's ratio of 2 was used in this example. As expected, the simple ratio with a factor of 2 may vary for nonspherical molecules. Under these assumptions, it can be concluded that the estimated value is an acceptable description of the surface molecules.

2.7.4 Other Surface Properties of Liquids

There are a variety of other surface chemical properties of liquid surfaces. This arises from the fact that liquids are stabilized by different forces. Since these are beyond the scope of this book, only a few important examples will be mentioned.

It is a common observation that if a stone is thrown into a lake it gives rise to waves. Liquid surfaces, for example on oceans or lakes, exhibit wave formation when strong winds are blowing over them. It is known that such waves are created by the wind energy being transposed to waves. Hence, mankind has tried to convert wave energy into other useful forms of energy sources. Both transverse capillary waves and longitudinal waves can deliver information about the elasticity and viscosity of surfaces, albeit on very different time scales. Rates of adsorption and desorption can also be deduced. Transverse capillary waves are usually generated with frequencies between 100 and 300 Hz. The generator is a hydrophobic knife edge situated in the surface and oscillating vertically, while the usual detector is a lightweight hydrophobic wire lying in the surface parallel to the generator edge. The generator and detector are usually close together (15–20 mm) so reflections set up a pattern of standing transverse waves. Optical detection, on the other hand, causes no interference to the generate pattern. The damping of capillary ripples arises primarily from the compression and expansion of the surface and the interaction between surface film and of propagation at the top of the path and in the reverse direction at the bottom (Adamson and Gast, 1997). This leads to compression and expansion of the surface. If a surface film is present, compression tends to lower the surface tension while expansion raises it. This generates a Marangoni flow, which opposes the wave motion and dampens it (Birdi, 2002, 2010a). Furthermore, if the material is soluble, the compression–expansion cycle will be accompanied with the hydrodynamic theories for capillary ripples. For longitudinal waves, a barrier (usually on the trough of a surface film) generates the length of the trough. If the surface tension changes in the given area of observation, it also has an effect on the surface waves.

2.7.5 Interfacial Tension of Liquid$_1$–Liquid$_2$

Oil and water do not mix; this is an everyday observation. The main reason being that oil is insoluble in water, and vice versa. Thus, the oil–water interface has interfacial surface forces. In this chapter, the methods that can be used to disperse oil in water (or vice versa) will be described. An analysis of the interfacial tension (IFT) that exists at any oil–water interface will be described. In the literature, the interfacial tension, γ_{AB}, between two liquids with γ_A and γ_B has been described in much detail (Adamson and Gast, 1997; Chattoraj and Birdi, 1984; Somasundaran, 2006). An empirical relation has been suggested (Antonow's rule) by which one can predict the surface tension γ_{AB}:

$$\gamma_{AB} = | \gamma_{A(B)} - \gamma_{B(A)} | \tag{2.52}$$

The prediction of γ_{AB} from this rule is approximate but has been useful in a large number of systems (such as alkanes:water), with some exceptions (such as water:butanol) (Table 2.3). For example:

$$\gamma_{water} = 72 \text{ mN/m (at 25°C)}$$
$$\gamma_{hexadecane} = 20 \text{ mN/m (at 25°C)}$$
$$\gamma_{water\text{-}hexadecane} = 72 - 20 = 52 \text{ mN/m (measured} = 50 \text{ mN/m)} \tag{2.53}$$

However, for general considerations one may only use it as a reliable guideline, and when exact data is not available. Antonow's rule can be understood in terms of a simple physical picture. There should be an adsorbed film or Gibbs monolayer of substance B (the one of lower surface tension) on the surface of liquid A. If we regard this film as having the properties of bulk liquid B, then $\gamma_{A(B)}$ is effectively the interfacial tension of a duplex surface and would be equal to $\gamma_{A(B)} + \gamma_{B(A)}$.

TABLE 2.3
Antonow's Rule and Interfacial Tension Data (mN/m)

Oil Phase	W(O)	O(W)	O/W	W(O)-O
Benzene	62	28	34	34
Chloroform	52	27	23	24
Ether	27	17	8	9
Toluene	64	28	36	36
n-Propylbenzene	68	29	39	40
n-Butylbenzene	69	29	41	40
Nitrobenzene	68	43	25	25
i-Pentanol	28	25	5	3
n-Heptanol	29	27	8	2
CS2	72	52	41	20
Metyleneiodide	72	51	46	22

Interfacial tension can be measured by different methods, depending on the characteristics of the system. The following methods can be applied:

- Wilhelmy plate method
- Drop weight method (can be also used for high pressure and temperature)
- Drop shape method (can be also used for high pressure and temperature)

The Wilhelmy plate is placed at the surface of the water, and the oil phase is added until the whole plate is covered by the latter. The apparatus must be calibrated with known IFT data, such as water–hexadecane (52 mN/m; 25°C) (Table 2.4).

TABLE 2.4
Interfacial Tensions (IFT) between Water and Organic Liquids (20°C)

Water/Organic Liquid	IFT (mN/m)
n-Hexane	51.0
n-Octane	50.8
CS2	48.0
CCl4	45.1
Br-C6 H5	38.1
C6 H6	35.0
NO2 -C6 H5	26.0
Ethyl ether	10.7
n-Decanol	10
n-Octanol	8.5
n-Hexanol	6.8
Aniline	5.9
n-Pentanol	4.4
Ethyl Acetate	2.9
Isobutanol	2.1
n-Butanol	1.6

The drop weight method is carried out by using a pump (or a syringe) to deliver the liquid phase into the oil phase (or vice versa, as one finds suitable). In the case of water, the water drops sink to the bottom of the oil phase. The weight of the drops is measured (by using an electrobalance) and IFT can be calculated. The accuracy can be very high by choosing the right kind of setup. The drop shape (pendant drop) is most convenient if small amounts of fluids are available, as well as if extreme temperatures and pressure are involved. Modern digital image analyses also make this method very easy to apply in extreme situations.

Appendix 2A: Effect of Temperature and Pressure on the Surface Tension of Liquids (Corresponding States Theory of Liquids)

Both in industry and research, a large data of substances can be manipulated, which can be systemized in order to predict and understand the systems. Accordingly, one is also interested in understanding the chemistry and the physics of liquid surfaces. It is thus important to be able to describe the interfacial forces of liquids as a function of temperature and pressure. This is most important in the case of oil recovery from reservoirs where oil is found at high temperatures and pressures. The magnitude of γ decreases almost linearly with temperature (t) (for most liquids) within a narrow range (Defay et al., 1966; Kuespert et al., 1995; Birdi, 2002, 2009):

$$\gamma_t = k_o (1 - k_1 t) \tag{2A.1}$$

where k_o is a constant. It was found that coefficient k_1 is approximately equal to the rate of decrease of density, ρ, of liquids with rise of temperature:

$$\rho_t = \rho_o (1 - k_1 t) \tag{2A.2}$$

where ρ_o is the value of density at t = 0 C, and the values of constant k_1 are different for different liquids. Furthermore, the value of γ was related to critical temperature (T_C).

The following equation relates surface tension of a liquid to the density of liquid, ρ_l, and vapor, ρ_v (Partington, 1951; Birdi, 1989, 2010a, 2010b):

$$\gamma/(\rho_l - \rho_v)^4 = C \tag{2A.3}$$

where the value of constant C is nonvariable only for organic liquids, while it is not constant for liquid metals. At the critical temperature, T_c, and critical pressure, P_c, a liquid and its vapor are identical, and the surface tension, γ, and total surface energy, like the energy of vaporization, must be zero (Birdi, 1997). At temperatures below the boiling point, which is 2/3 T_c, the total surface energy and the energy of evaporation are nearly constant. The variation in surface tension, γ, with temperature is given in Figure 2.21 for different liquids.

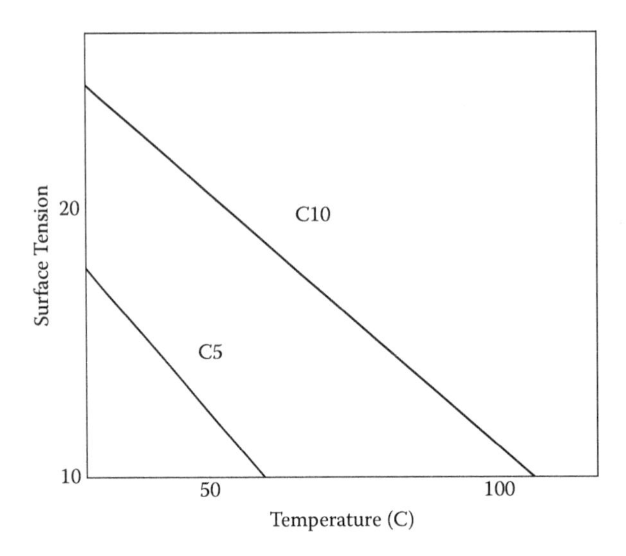

FIGURE 2.21 Variation of γ with temperature of different alkanes (C5: n-pentane; C10: n-decane).

These data clearly show that the variation of γ with temperature is a very characteristic physical property. This observation becomes even more important when it is considered that the sensitivity of γ measurements can be as high as approximately ±0.001 dyne/cm(= mN/m) (as described in detail next). The change in γ with temperature in the case of mixtures would thus be dependent on the composition. For example the variation of γ of the system: CH_4 + Hexane, is given as follows:

$$\gamma_{(CH4+\,hexane)} = 0.64 + 17.85\ x_{hexane} \tag{2A.4}$$

This relation shows that hexane increases the magnitude of γ of the mixture. This means that in any such mixture the magnitude will increase when the additive has higher γ than the other component. It is seen (Equation 2A.4) that actually by measuring γ for such system one can estimate the concentration of hexane in CH_4. This is of much interest in oil reservoir engineering operations where CH_4 is found in crude oil. The addition of a gas to a liquid always decreases the value of γ in the mixture. It is well known that the corresponding states theory can provide much useful information about the thermodynamics and transport properties of fluids. For example, the most useful two-parameter empirical expression that relates the surface tension, γ, to the critical temperature, T_c, is given as:

$$\gamma = k_o\ (1 - T/T_c)^{k1} \tag{2A.5}$$

where k_o and k_1 are constants. The magnitude of constant k_1 has been reported as 3/2, although experiments indicated that k_1 = 1.23 or 11/9 (Birdi, 1997, 2002). Data for some liquids has also reported the value of k_1 to be between 6/5 and 5/4.

It has been found that constant k_o was proportional to $T_c^{1/3} P_c^{2/3}$. The relation in Equation (A.5) when fitted to the surface tension, γ, data of liquid CH_4, has been found to give the following relationship:

$$\gamma_{CH4} = 40.52\ (1 - T/190.55)^{1.287} \tag{2A.6}$$

where T_c = 190.55K. This equation has been found to fit the γ data for liquid methane from 91 to –190 K, with an accuracy of ±0.5 mN/m. In a recent study, the γ versus T data on n-alkanes, from n-pentane to n-hexadecane, were analyzed (Birdi, 1997, 2010a). The constants k_o (between 52 and 58) and k_1 (between 1.2 and 1.5) were found to be dependent on the number of carbon atoms, n_C, and since T_c is also dependent on n_C, the expression for all the different alkanes that individually were fit to Equation (2A.5) gave rise to a general equation where γ was a function of C_n and T as follows (Birdi, 1997). The estimated values of different n-alkanes were found to agree with the measured data within a few percents: γ for n-C18 H38, at 100°C, was 21.6 mN/m, from both measured and calculated values. This agreement shows that the surface tension data on n-alkanes fits the corresponding state equation very satisfactorily. It is worth mentioning that the equation for the data on γ versus T, for polar (and associating) molecules like water and alcohol, when analyzed by Equation (2A.5), gives magnitudes of k_o and k_1, which are significantly different than those found for nonpolar molecules such as alkanes (Birdi, 1997).

In the following, calculated values of γ for different alkanes are given based upon the analyses using Equation (2A.5).

Calculated[a] γ and Measured Values of Different n-Alkanes at Various Temperatures

-n-Alkane	Temperature/°C	Measured	Calculated
C5	0	18.23	18.25
	50	12.91	12.8
C6	0	20.45	20.40
	60	14.31	14.3
C7	30	19.16	19.17
	80	14.31	14.26
C9	0	24.76	24.70
	50	19.97	20.05
	100	15.41	15.4
C14	10	27.47	27.4
	100	19.66	19.60
C16	50	24.90	24.90
C18	30	27.50	27.50
	100	21.58	21.60

[a] Calculations from Birdi, K. S., Editor, *Handbook of Surface & Colloid Chemistry*, CRC Press, Boca Raton, FL, 1997.

The variation of γ for water with temperature (t/C) is given as (Cini et al., 1972; Birdi, 1997):

$$\gamma_{H2O} = 75.668 - 0.1396\, t - 0.2885\, 10^{-3}\, t^2 \quad (2A.7)$$

This data is useful since water is used as a calibration liquid in many surface tension studies.

The surface entropy (S_s) corresponding to Equation (2A.5) is:

$$S_S = -d\,\gamma/dT$$

$$= k_1\, k_o\, (1 - T/T_c)^{k1} - 1/(T_c) \quad (2A.8)$$

and the corresponding surface enthalpy, h_s:

$$h_s = g_s - T\, S_s$$

$$= -T\, (d\,\gamma/dT)$$

$$= k_o\, (1 - T/T_c)^{k1-1}\, (1 + (k_1 - 1)\, T/T_c) \quad (2A.9)$$

The reason heat is absorbed on expansion of a surface is that the molecules must be transferred from the interior against the inward attractive force to form the new surface. In this process, the motion of the molecules is retarded by this inward attraction, so that the temperature of the surface layers is lower than that of the interior, unless heat is supplied from the outside. The same is true when a gas molecule adsorbs on a solid surface.

- Oil reservoirs are found at 100 to 200 atm pressure and high temperatures (80°C).
- High-pressure technologies.
- Car tires exert high pressure on the roads.
- Teeth in the mouth exert considerable high pressure.
- Shoes (e.g., soles) are exposed to high pressures.
- Building structures.

The following relationship relates γ to density (Birdi, 1997):

$$\gamma\, (M/\rho)^{2/3} = k\, (T_c - T - 6) \quad (2A.10)$$

where M is the molecular weight, ρ is the density, (M/ρ = molar volume). The quantity ($\gamma\, (M/\rho)^{2/3}$) is called the molecular surface energy. It is important to note the correction term 6 on the right-hand side. This is the same as found for n-alkanes and n-alkenes in the estimation of T_c from γ versus temperature data (Birdi, 1997). Surface tension and temperature relationships are useful in the oil industry.

3 Surfactant (Soaps and Detergents) Solutions: Essential Surface Properties

3.1 INTRODUCTION

In the history of mankind, the washing process is probably one of the oldest known systems that relates to surface chemistry. As known from experience, any physical property of a liquid will change when a substance (called a *solute)* is dissolved in it. Of course, the change may be small or large, depending on the concentration and other parameters. Accordingly, the magnitude of surface tension of a liquid will change when a solute is dissolved in it. This will be expected from physicochemical considerations. It also becomes apparent that if the surface tension of water could be manipulated, then many application areas would be drastically affected. In this chapter some important *surface active substances* with such properties will be described. These are some specific substances that are used to change the surface tension of water in order to apply this characteristic for some useful purpose in everyday life.

The magnitude of surface tension change will depend on the concentration and on the solute added. In some cases the surface tension, γ, of the solution increases (such as NaCl and other salts). The change in surface tension may be small (per mole of added solute) (as in the case of inorganic salts) or large (as in the case of such molecules as ethanol or other soap-like molecules) (Figure 3.1).

Surface tension of typical inorganic salt–water solutions (20°C):

Pure water, 72.75 mN/m
NaCl (0.1 mol/liter), 72.92 mN/m
Nal (0.93 mol/liter), 74.4 mN/m
NaCl (4.43 mol/liter), 82.55 mN/m
KCl (0.93 mol/liter), 74.15 mN/m
HCl (0.97 mol/liter), 72.45 mN/m

KCl gives a larger increase than NaCl. This indicates that the degree of adsorption of KCl at the surface is larger than that of NaCl. Further, experiments have

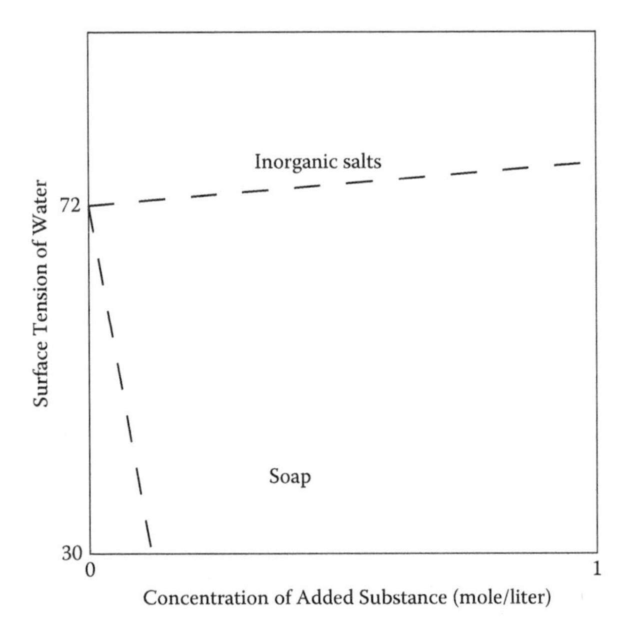

FIGURE 3.1 Change in surface tension of water on addition of inorganic salt (for example, NaCl or KCl) or a soap.

shown that the solutions of salts exhibit surface charges. This arises from the fact that the number of positive ions and negative ions is not equal (contrary to the bulk phase). In other words, it has been found that the surface potential of KCl solution is lesser than in the case of NaCl. From this it has been concluded that K^+ and Cl^- ions adsorb at the surface almost to the same degree. These phenomena have very significant consequences in systems where these ions are present in connection with surfaces. One example is that in biological cells the role of Na^+ and K^+ is significantly different.

Change in γ with the addition of a solute (equal molar concentration):

Inorganic salt, minor change (increase) in γ
Ethanol or similar, small change (decrease) in γ
Soap (similar), large change (decrease) in γ

Some typical surface tension data of different solutions are given next.

Surface Tension (mN/m)	72	50	40	30	22
Surfactant					
C12H25SO4Na	0	0.0008	0.003	0.008	—
Ethanol	0	10	20	40	100

This shows that to reduce the value of γ in water from 72 to 30 mN/m, 0.005 moles of sodium dodecyl sulfate (SDS) or 40% ethanol would be needed. Of course, these two solutions cannot be used for same application based on their similar magnitude of γ.

There are special substances called *soaps* or *detergents* or *surfactants* that exhibit unique physicochemical properties (Rosen, 2004; Birdi, 2009, 2010a). The most significant structure of these molecules is due to the presence of a *hydrophobic* (alkyl group) and a *hydrophilic* (polar group, such as -OH; $-CH_2CH_2O^-$; -COONa; $-SO_3Na$; $-SO_4Na$; $-CH_33N-$).

The different polar groups are:

Ionic groups
  Negatively charged, *anionic*
    -COONa
    SO_3Na
    $-SO_4Na$
  Positively charged, *cationic*
    $-(N)(CH_3)_4Br$
  Amphoteric
    $-(N)(CH_3)_2-CH_2-COONa$

Nonionic groups
  -CH2CH2OCH2CH2OCH2CH2OH
  -(CH2CH2OCH2CH2O)x(CH2CH2CH2O)yOH

Accordingly, these substances are also called *amphiphile* (meaning "two kinds") (i.e., alkyl part and the polar group):

CCCCCCCCCCCCCCCCCCCC-**O**
ALKYL GROUP(CCCCCCC-)- POLAR GROUP(-**O**) =
AMPHIPHILE
(CH3CH2CH2CH2CH2CH2CH2CH2CH2)-POLAR

In drawings, the alkyl group is depicted as — — — —, whereas the polar group is depicted as 0.

For instance, surfactants dissolve in water and give rise to low surface tension (even at very low concentrations [few grams per liter or 1 to 100 mmol/liter]) of the solution, therefore these substances are also called surface active molecules (*surface active agents* or substances). On the other hand, most inorganic salts increase surface tension of water. All surfactant molecules are amphiphilic, which means these molecules exhibit hydrophilic and hydrophobic properties. Ethanol reduces the surface tension of water, over a few moles per liter will be needed to obtain the same reduction as when using a few millimoles of surface active agents. As expected, if ethanol (with γ of 22 mN/m) is added to water (with γ of 72 mN/m), then the magnitude of the mixed solution will decrease. This is always the case in mixtures. Further, if one dissolves a gas in a liquid, the magnitude of γ will always decrease. This also means that if one dissolves a gas in a liquid, then the magnitude of the mixture will always decrease. In fact, the amount of gas that is dissolved can be estimated by measuring the decrease of γ in such mixtures.

However, the value of γ of surfactant solutions decreases to 30 mN/m with surfactant concentration around mmol/L (range of 1 to 10 g/L). Soaps have been used by mankind for many centuries. In biology, there is a whole range of natural amphiphile molecules (bile salts, fatty acids, cholesterol and other related molecules, phospholipids). In fact, many important biological structures and functions are based on amphiphile molecules.

It is important to mention that surfactants are one of the most important types of substances that play an essential role in everyday life. Moreover, many surfactants in nature (such as bile acids in the stomach, which basically behave exactly the same way as the man-made surface active agent) also exist. Proteins (which are large molecules, with molecular weights from 6000 to millions) also decrease γ when dissolved in water. Different proteins decrease dependent on the amino acid composition. Soaps and surfactants are molecules that are characterized as amphiphiles. A part of the amphiphile likes oil or hydrophobic (lipophilic = likes fat) (Tanford, 1980), while the other part likes water or hydrophilic (also called lipophobic). The balance between these two parts, hydrophilic–lipophilic, is called HLB. The latter quantity can be estimated by experimental means and theoretical analyses allow one to estimate its value (Adamson and Gast, 1997; Birdi, 2009). HLB values are applied in the emulsion industry (Hansen, 2007; Birdi, 2009).

Soap molecules are made by reacting fats with strong alkaline solutions (this process is called saponification). In water solutions the soap molecule, $C_nH_{2n+1}COONa$ (with n greater than 12 to 22), dissociates at high pH into $RCOO^-$ and Na^+ ions. It has been found that the magnitude of n must be 12 or more for effective results. Originally, man used the soaps as produced from fats (and lard). However, many decades ago synthetic surfactants were made (as a by-product from oil refinery products) to apply to special industrial applications, such as cleaning and washing processes. A great variety of surfactants were synthesized from oil by-products, especially $C_{12}H_{25}C_6H_4SO_3Na$ (sodium dodecyl benzene sulfonates), were used in detergents. Later, these were replaced by sodium dodecyl sulfates or sulfonates, because the sodium dodecyl benzenes were not biodegradable. This means that the bacteria in the sewage plants were not able to degrade the alkyl group. The alkyl sulfonates were degraded to alkyl hydroxyls and alkyl aldehydes and later to CO_2. This problem has been solved after one uses the biodegradable alkyl sulfates. In many applications one found it necessary to employ surfactants that were nonionic. A variety of nonionic surfactants have been synthesized and tailor-made surfactants that suit one particular application can be obtained. Further, since nonionic detergents do not exhibit any charges means that applications where this property is essential, for example, critical micelle concentration (CMC) of nonionics does not change.

For example, nonionic detergents as used in washing clothes are much different in structure and properties than those used in dishwashing machines. In washing machines, foam is crucial as it helps to keep dirt away from the clothes once it has been removed. On the other hand, foam is not needed in machine dishwashing. This arises from the fact that foam will hinder the mechanical effect of the dishwashing process. However, surface tension (which means wetting and other properties at the interface) needs to be low for decreasing the contact angle and as well as to remove fats (through detergent action). There are tailor-made nonionics that satisfy this criteria. Another important property of surface active agents is that most of the ionic detergents cannot be used in conjunction with seawater (due to its high content of Ca and Mg salts). This is due to the fact that Ca salts and Mg salts of ionic detergents are highly insoluble in water. Therefore, special detergents have been used to resolve this problem.

3.2 SURFACE TENSION PROPERTIES OF AQUEOUS SOLUTIONS

The surface tension of any pure liquid (water or organic liquid) will change when another substance (solute) is dissolved. The change (increase or decrease) in surface tension will depend on the characteristics of the solute added. The surface tension of water increases (in general) when inorganic salts (such as NaCl, KCl, and Na_2SO_4) are added (Figure 3.1), while its value decreases when organic substances are dissolved (ethanol, methanol, fatty acids, soaps, detergents) (Figure 3.2).

The surface tension of water increases from 72 mN/m to 73 mN/m when 1 M NaCl is added. On the other hand, the magnitude of surface tension decreases from 72 mN/m to 39 mN/m when only 0.008 M (0.008 M × 288 = 2.3 g/Liter) SDS (mol. Wt. = 288) is dissolved. It thus becomes obvious that in all those systems where surface tension plays an important role, the additives will become significant in these systems. The data of n-butanol solutions in water are shown to decrease from 72 mN/m (pure water) to 50 mN/m in 200 mmol/Liter (Figure 3.3).

The magnitude of surface tension changes slowly in the case of methanol as compared to detergent solutions. The methanol–water mixtures gave the following surface tension data (20°C):

%w methanol	0	10	25	50	80	90	100
γ (mN/m)	72	59	46	35	27	25	22.7

It is important to have an understanding about the change in surface tension of water as a function of molecular structure of solute. The surface tension data in the case of homologous series of alcohols and acids show some simple relation to the alkyl chain length (Figure 3.4). It is noticed that each addition of -CH2- group gives a value of concentration and surface tension such that the value of concentration is lower by about a factor of 3.

However, it must be mentioned that such dependence in the case of nonlinear alkyl chains will be different. The effective -CH2- increase in the case of a nonlinear

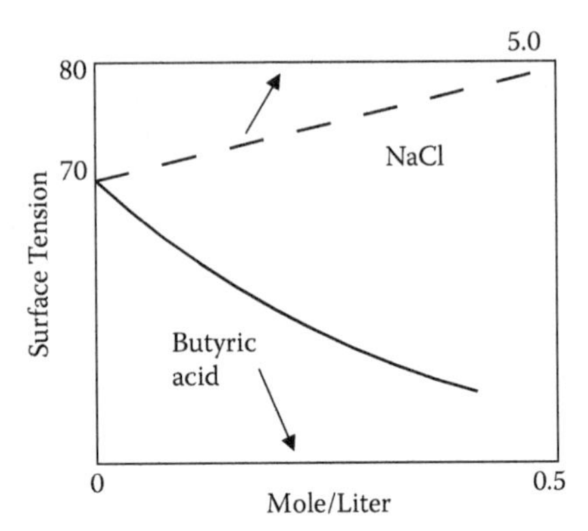

FIGURE 3.2 A change in the surface tension of water as a function of added solutes (inorganic salts, surface active agents) (NaCl, butyric acid).

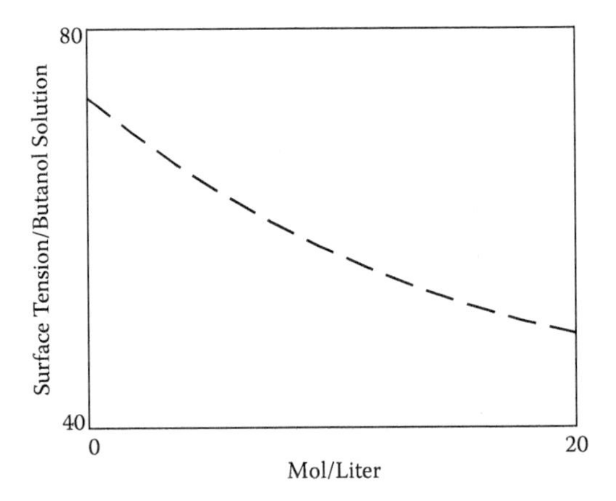

FIGURE 3.3 Surface tension plot of n-butanol solutions.

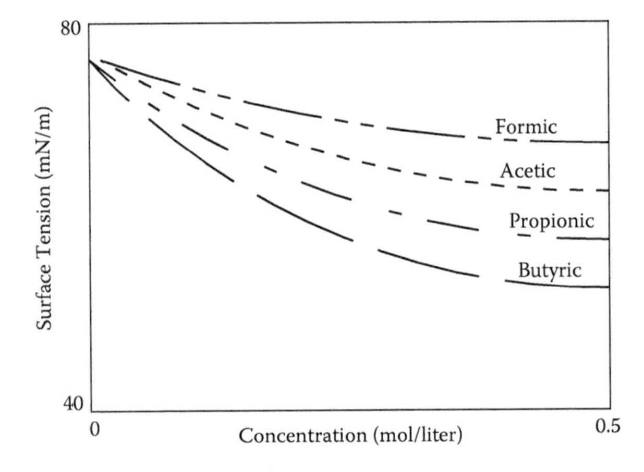

FIGURE 3.4 Surface tension data of some homologous series of short-chain acids in water.

chain will be lesser (ca. 50%) than in the case of a linear alkyl chain. The tertiary -CH2- group effect would be even lesser. In general, though, one will expect that the change in surface tension per mole substance will increase with any increase in the hydrocarbon group of the amphiphile.

The effect of chain length on surface tension arises from the fact that as the hydrophobicity increases with each -CH2- group, the amphiphile molecule adsorbs more at the surface. This will thus be a general trend also in more complicated molecules, such as in proteins and other polymers.

In proteins, the amphiphilic property arises from the different kinds of amino acids (25 different amino acids). Some amino acids have lipophilic groups (such as phenylalanine, valine, and leucine), while others have hydrophilic groups (such as glycine and aspartic acid).

In fact, from surface tension measurements, some proteins are considerably more hydrophobic (such as hemoglobin) than others (such as bovine serum albumin or ovalbumin). These properties of proteins have been extensively investigated and these data have been related to biological functions (Chattoraj and Birdi, 1984; Birdi, 1999).

3.2.1 Surface Active Substances (Amphiphiles)

All molecules that when dissolved in water reduce surface tension are called surface active substances (such as soaps, surfactants, detergents). This means that surface active substances adsorb at the surface and reduce surface tension. The same will happen if a surface active substance is added to a system of the oil–water. The interfacial tension of the oil–water interface will be reduced accordingly. Inorganic salts on the other hand, increase the surface tension of water.

Surfactants exhibit surface activity, which means these molecules will adsorb preferentially at interfaces:

- Air–water
- Oil–water
- Solid–water

The magnitude of surface tension is reduced since the hydrophobic (alkyl chain or group) is energetically more attracted to the surface than being surrounded by water molecules inside the bulk aqueous phase. Figure 3.5 shows the monolayer formation of the surface active substance at high bulk concentration. Since the close-packed surface active substance at the surface looks like alkane, it would be thus expected that the surface tension of surface active substances solution would decrease from 72 mN/m (surface tension of pure water) to alkane-like surface tension (close to 25 mN/m).

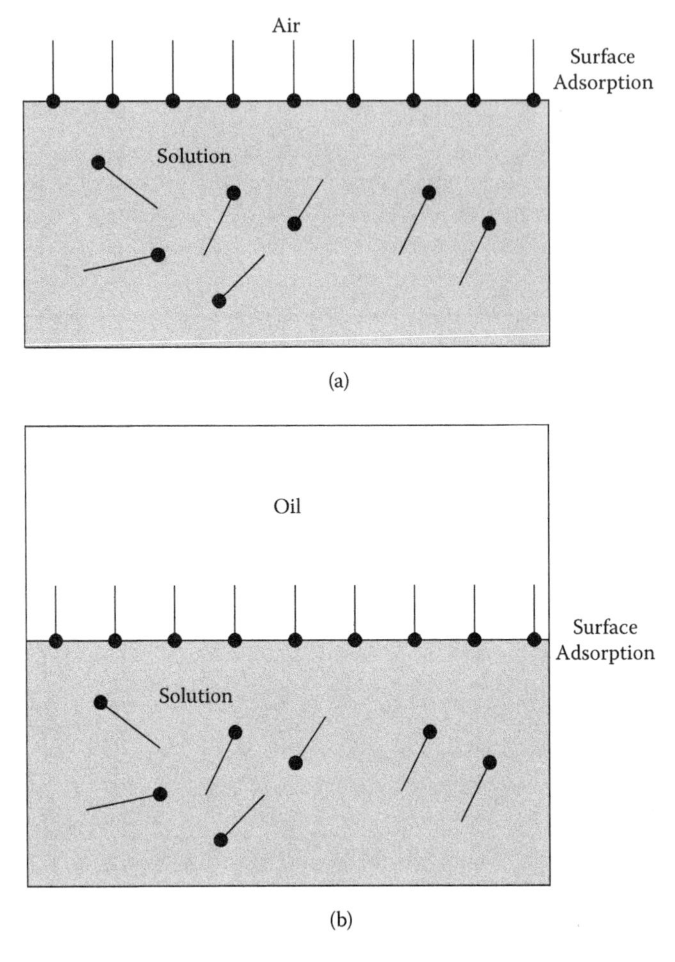

FIGURE 3.5 The orientation of soap (surface active substance) at (a) the surface of water and (b) the oil–water interface (alkyl group: — —; polar group: 0).

The orientation of surface molecules at the interface will be dependent on the system. This is shown as follows:

- Air–water—Polar part toward water and hydrocarbon part toward air
- Oil–water—Polar part toward water and hydrocarbon part toward oil
- Solid–water—Polar part toward water and hydrocarbon part toward solid (in general)

3.2.2 Aqueous Solution of Surfactants

The solution properties of the various surfactants in water are very unique and complex in many aspects, as compared to such solutes as NaCl or ethanol. However, in the following a very simplified but useful and practical description will be given. For more detailed aspects, the reader is advised to look up the relevant references (Tanford, 1980; Birdi, 2002; Rosen 2004). These solubility characteristics thus require special description, which is given here. The solubility of charged and noncharged surfactants is very different, especially with regard to the effect of temperature and salts (such as NaCl). These characteristics are important when these substances need to be applied in diverse systems. For instance, one cannot use the same soap molecule at sea as on the land. The main reason being that higher concentrations of salts (such as Ca^{++} and Mg^{++}) as found in seawater affect the foaming and solubility characteristics of major surface active substances. For similar reasons, a nonionic detergent for shampoos cannot be used (only anionic detergents are used). Therefore, tailor-made surface active agents have been devised by the industry to meet these specific demands. In fact, the whole soap industry develops detergents designed for each specific system. Industrial application research in this area is very extensive and protected by patents.

3.2.3 Solubility Characteristics of Surfactants in Water

The solubility characteristics of any substance are very important information that one must investigate. In the present case, one must have the precise information about the solubility and temperature characteristics of the surfactant. The solubility characteristics of surfactants (in water) is one of the most studied phenomena. Even though the molecular structures of surfactants are rather simple, the solubility in water is rather complex as compared with other amphiphiles, such as long chain alcohols. The solubility in water will be dependent on the alkyl group and as well as on the polar group. This is easily seen from the fact that the alkyl groups will behave mostly as alkanes. However, it is also found that the solubility of surfactants is also dependent on the presence (or absence) of charge on the polar group. The ionic surfactants exhibit different solubility characteristics than the nonionic surfactants, with regard to dependence on the temperature. In fact, in all industrial applications of surface active substance the solubility parameter is one of the most important criteria. This characteristic is the determining factor about which surface active substance to be used in a given system. For example, the surface active substance needed for household washing detergents will be different from the one needed where seawater is used for washing. This arises from the fact that seawater has different salts (Ca and Mg) that are not found in high concentrations in normal household conditions. The hydrophobic alkyl part exhibits solubility in water that has been related to a surface tension model of the *cavity* (Appendix 3B).

3.2.3.1 Ionic Surfactants

The solubility of all ionic surfactants (both anionics, that is, negatively charged and cationics, that is, positively charged) is low at low temperatures but at a specific temperature the solubility suddenly increases (Figure 3.6). For instance, the solubility of SDS at 15°C is about 2 g/L. This temperature is called the Krafft point (KP). The Krafft point (or temperature) can be obtained by cooling an anionic surfactant solution (ca. 0.5 molar) from a high to a lower temperature until cloudiness appears sharply. The KP is not very sharp in the case of impure surfactants as generally found in industry.

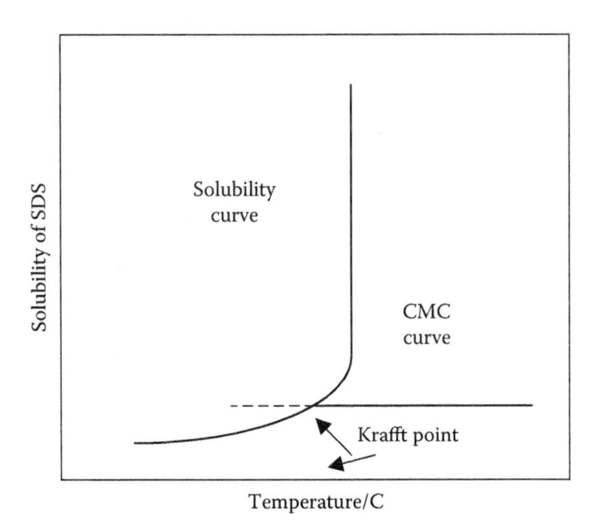

FIGURE 3.6 The solubility (the Kraftt point, KP) of ionic (anionic or cationic) surfactants in water (as a function of temperature).

It has been found that the magnitude of solubility near the KP is almost equal to the CMC. The magnitude of KP is dependent on the chain length of the alkyl chain for a homologous series of nonionic detergents (Figure 3.7).

The linear dependence of KP on the alkyl chain length is very clear. KP for C12 sulfate is 21°C, and it is 34°C for C14 sulfate. It can be concluded that KP increases by approximately 10°C per CH_2 group. Since no micelles can be formed below the KP, it is important to keep this information in mind when using any anionic detergent. Therefore, the effect of various parameters on the KP needs to be considered in the

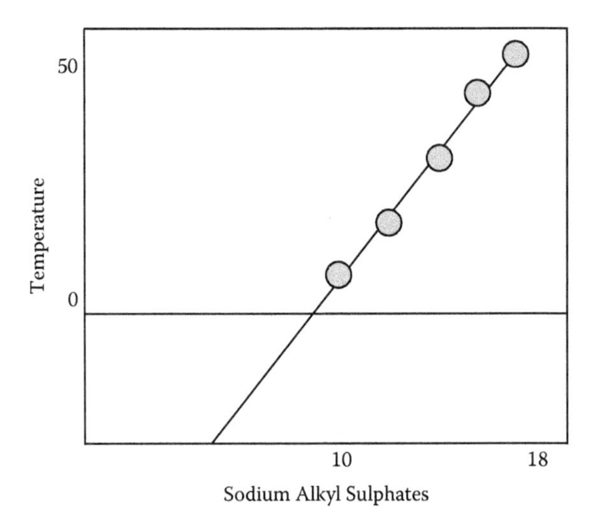

FIGURE 3.7 A variation of KP with a chain length of sodium alkyl sulfates.

case of ionic surfactants. For example, KP increases with alkyl chain length and KP decreases if a lower chain surfactant is mixed with a longer chain surfactant.

3.2.3.2 Nonionic Surfactants

Molecules with ionic charges behave differently than those with no charge, as regards solubility in water (besides other properties). The solubility of nonionic surfactants in water is completely different than those of charged surfactants (especially with regard to the effect of temperature). The solubility of nonionic surfactant is high at low temperatures but it decreases abruptly at a specific temperature, called the *cloud point* (CP) (Figure 3.8). This means that nonionic detergents will not be suitable if used above the cloud point temperature. The solubility of such detergent molecules in water arises from the hydrogen bond formation between the hydroxyl (-OH) and ethoxy groups (-CH2CH2O-) and the water molecules. At high temperatures the degree of hydrogen bonding gets weaker (due to high molecular vibrations) and thus the nonionic detergents become insoluble at the cloud point. The name cloud point is at the temperatures when the solution becomes cloudy (because a new phase with surfactant rich concentration is formed). The solution separates into two phases with a rich water phase and low concentration of nonionic surfactant. The rich nonionic detergent phase is found to consist of low water content. Experiments have shown that there are roughly four molecules of bound water per ethylene oxide group (-CH2CH2O-) (Birdi, 2009).

Thus, it must be noted that when a surfactant is needed for any application the solubility characteristics must be considered in addition to other properties, which should conform to the experimental conditions. Thus, the surfactants that are available in the industry are characterized by their area of application. In fact, the

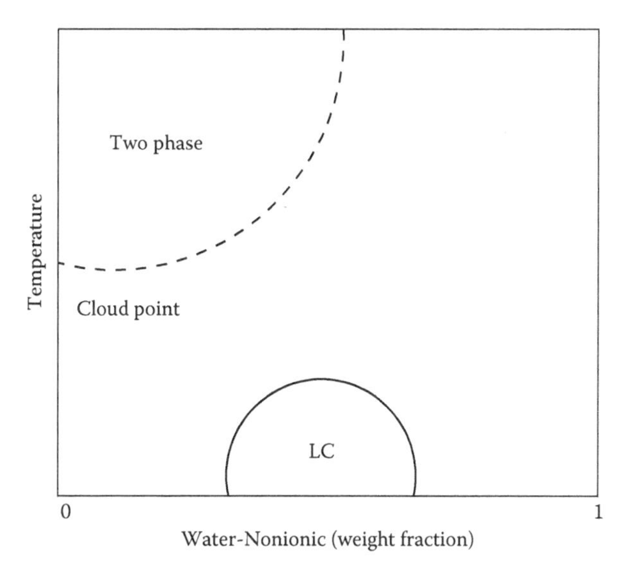

FIGURE 3.8 Solubility of a nonionic surfactant in water (cloud point, CP) (dependent on temperature).

detergent manufacturer is in constant collaboration with the washing industry, and tailor-made surfactants are commonly developed in collaboration.

Anionic surfactants are used in different areas, whereas cationics are used completely different. For instance, anionic surfactants are used for shampoos and washing, whereas cationics are used for hair conditioners. The hair has a negative charged surface and thus cationic strongly adsorb at the surface and leave a smooth surface. These detergents are used in the so-called *hair conditioner formulations*. This is because the charged end is oriented toward the hair surface and the alkyl group is pointing away (as depicted next: dashes indicate alkyl chains; ellipses indicate polar groups).

Cationic detergent (positive charge) + Hair (negative charge):

— — — -alkyl group...polar group/(+)hair(–)

This conditioner process is based on interaction of a positive charged molecule (cationic detergent) with a negative charged substance (hair). This imparts hydrophobicity to hair making it feel soft and look fluffy. Similar treatments are used in other industries to change the behavior of negatively charged substances after treatment with positively charged molecules. The most important is observed in biology, where most of the drugs are positively charged, since this allows a stronger interaction to biological cells (which are generally negatively charged).

3.2.4 Micelle Formation of Surfactants in Aqueous Media (Critical Micelle Concentration)

The solution properties of ordinary salts, such as NaCl, in water are rather simple. One can dissolve at a given temperature a specific amount of NaCl giving a saturated solution, approximately 5 mole/liter. Similarly, the solution characteristics of methanol or ethanol in water are also simple and straightforward. However, the solution behavior of surfactant molecules in water is much more complex. In addition to the effect on surface tension, the solution behavior is dependent on the charge of the surfactant. The surfactant aqueous solutions manifest two major forces that determine the solution behavior. The alkyl part being hydrophobic would tend to separate out as a distinct phase while the polar part tends to stay in solution. The difference between these two opposing forces thus determines the solution properties. The factors that have to be considered are:

- The alky group and water
- The interaction of the alkyl hydrocarbon groups with themselves
- The solvation (through hydrogen bonding and hydration with water) of the polar groups
- Interactions between the solvated polar groups

According to the rules of physical chemistry science, micelle formation is a very intriguing system. Accordingly, there is a vast amount of published literature describing the accurate state of phases in such systems (Tanford, 1980; Birdi, 2002, 2010a). Below CMC, the detergent molecules are present as single monomers. Above CMC, one will have that monomers, C_{mono}, are in equilibrium with micelles, C_{mice}. The physical chemistry of such an equilibrium is of great interest in the literature. The micelle with the aggregation number, N_{ag}, is formed from monomers:

$$N_{ag} \text{ monomer} = \text{Micelle} \tag{3.1}$$

N_{ag} monomers that were surrounded by water aggregate together, above CMC, and form a micelle. In this process, the alkyl chains have transferred from the water phase to an alkane-like micelle interior. This occurs because the alkyl part is at a lower energy in micelle than in the water phase (as shown later). The aggregation process is stepwise. This is based on the fact that some surfactants, such as cholates, form micelles with few aggregation numbers, while SDS can form micelles varying from 100 to 1000 aggregation numbers.

The surfactant molecule forms a micellar aggregate at concentrations higher than CMC, because it moves from the water phase (higher energy) to the micelle phase (lower energy). The micelle reaches an equilibrium after a certain number of monomers have formed a micelle. This means that there are both attractive and opposing forces involved in this process. Otherwise, one would expect very large aggregates if there was only attractive force involved. This would mean phase separation, that is, two phases: one water-rich phase and another surfactant-rich phase. Thus aggregation is a specific property where instead of phase separation molecules are able to form small aggregates, micelles (not visible to naked eye), and form very stable micellar solutions.

Thus, one can write the standard free energy of a micelle formation, ΔG°_{mice}, as:

$$\Delta G^\circ_{mice} = \text{Attractive forces} + \text{Opposing forces} \tag{3.2}$$

If there are only attractive forces present (as in alkanes), then solubility in water will not be observed. The attractive forces are associated with the hydrophobic interactions between the alkyl part (alkyl–alkyl chain attraction) of the surfactant molecule, $\Delta G_{hydrphobic}$. The opposing forces arise from the polar part (charge–charge repulsion; polar group–hydration), ΔG_{polar}. These forces are of opposite signs. The attractive forces would lead to larger aggregates. The opposing forces would hinder

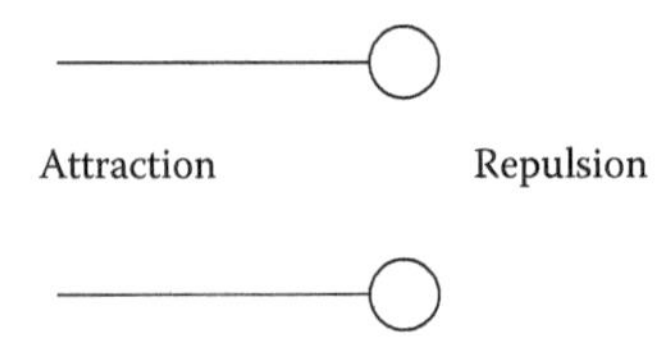

FIGURE 3.9 Attraction and repulsion between surfactant molecules.

the aggregation (Figure 3.9). A micelle with a definite aggregation number is where the value of ΔG°_{mice} is zero (Figure 3.10). Hence, we can write for ΔG°_{mice}:

$$\Delta G^\circ_{mice} = \Delta G_{hydrophobic} + \Delta G_{polar} \tag{3.3}$$

The standard free energy of micelle formation will be:

$$\begin{aligned}\Delta G^\circ_{mice} &= \mu^\circ_{mice} - \mu^\circ_{mono} \\ &= RT \ln (C_{mice}/C_{mono})\end{aligned} \tag{3.4}$$

At CMC, one may neglect C_{mice}, which leads to:

$$\Delta G^\circ_{mice} \approx R\,T \ln (CMC) \tag{3.5}$$

This relation is valid for nonionic surfactants but will be modified in the case of ionic surfactants (as shown later). This equilibrium shows that if we dilute the system then micelles will break down to monomers to achieve equilibrium. This is a simple equilibrium for a nonionic surfactant. In the case of ionic surfactants there will be

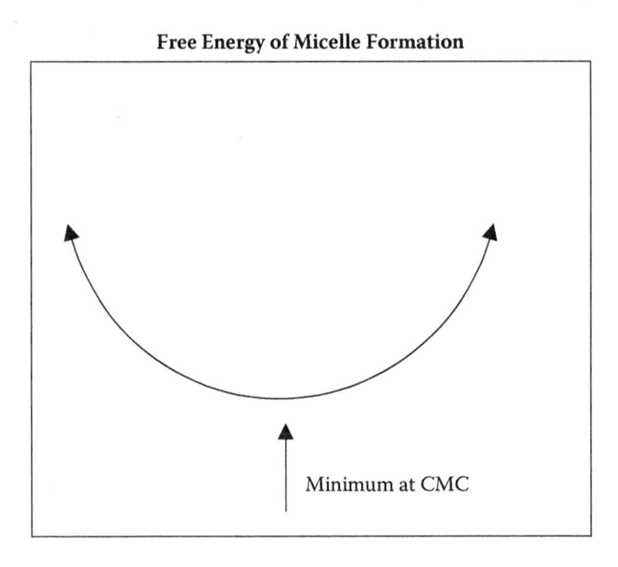

FIGURE 3.10 Attraction forces between alkyl chains and repulsion forces between polar groups gives minimum energy in the system at CMC.

charged species present in the solution. In the case of ionic surfactant, such as SDS, the micelle with aggregation number, N_{SD-}, will consist of counterions, C_{S+}:

$$N_{SD-} \text{ ionic surfactant monomers} + C_{S+} \text{ counter} = \text{Micelle with charge } (N_{SD-} - C_{S+}) \quad (3.6)$$

Since N will be larger than S^+, all anionic surfactants are negatively charged. Similarly, the cationic micelles will be positively charged. For instance, for CTAB, we have the following equilibria in micellar solutions: CTAB dissociates into CTA^+ and Br^- ions.

The micelle with N_{CTA+} monomers will have C_{Br-} counterions. The positive charge of the micelle will be the sum of positive and negative ions ($N_{CTA+} - C_{Br-}$). The actual concentration will vary with each species with the total detergent concentration. The case of SDS is shown in Figure 3.11a. The change in surface tension is given in Figure 3.11b.

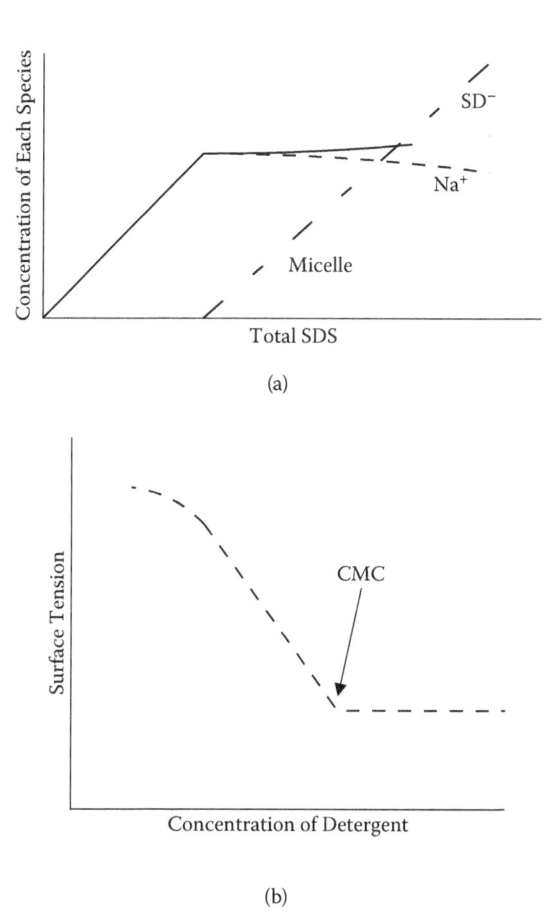

FIGURE 3.11 (a) Variation of concentration of different ionic species for SDS solutions (Na^+, SD^-, $SDS_{micelle}$); (b) change in surface tension of a detergent solution with concentration.

The surface tension curve (Figure 3.11b) is typical for all kinds of detergents. Below CMC the SDS molecules in water are found to dissociate into SD^- and Na^+ ions. Conductivity measurements show:

1. That SDS behaves as a strong salt and SD^- and Na^+ ions are formed (same as one observes for NaCl)
2. That a break on the plot is observed at SDS concentration equal to the CMC. This clearly shows that the number of ions decreases with concentration. The latter indicates that some ions (in the present case cations, Na^+) are partially bound to the SDS-micelles, which results in change in the slope of the conductivity of the solution. The same behavior is observed for other ionic detergents, such as cationic (CTAB) surfactants. The change in surface tension also shows a break at CMC.

At CMC, micelles (aggregates of SD^- with some counterions, Na^+) are formed and some Na^+ ions are bound to these, which is also observed from conductivity data. In fact, these data analyses have shown that approximately 70% Na^+ ions are bound to SD^- ions in the micelle. The surface charge (negative charge) was estimated from conductivity measurements (Birdi, 2002). Therefore, the concentration of Na^+ will be higher than SD^- ions after CMC. A large number of reports are found in literature where the transition from monomer phase (before CMC) to micellar phase (after CMC) has been extensively analyzed.

The same is true in the case of cationic surfactants. In the case of CTAB solutions, one thus has CTA^+ and Br^- ions below CMC. Above CMC, there are additional $CTAB^+$ micelles. In these systems the counterion is Br^-. It is important to note that due to these differences, the two systems are completely divergent from each other with regard to the areas of application (for example, CTAB cannot be used for washing clothes).

3.2.4.1 Analyses of CMC of Surfactants

The magnitude of CMC will be dependent on different factors. These will be both dependent on the alkyl part and the polar part. One can study this by varying the two parts in a series of suitable molecules. Further, the interaction of the detergent with the solvent will also have an effect on the CMC. Let us analyze the effect of the alkyl chain length.

It has been found that CMC decreases with increasing alkyl chain length. This indicates that as the solubility in water of the alkyl part decreases, the CMC also decreases. In linear chain Na-alkyl sulfate detergents, the following simple relationship has been found:

$$\ln (\mathrm{CMC}) = k_1 - k_2 (C_{alkyl}) \tag{3.7}$$

where k_1 and k_2 are constants, and C_{alkyl} is the number of carbon atoms in the alkyl chain.

The CMC will change if the additive has an effect on the monomer–micelle equilibrium. It will also change if the additive changes the detergent solubility. The CMC of all ionic surfactants will decrease if co-ions are added (Figure 3.12). However, nonionic surfactants show very little change in CMC with the addition of salts. This

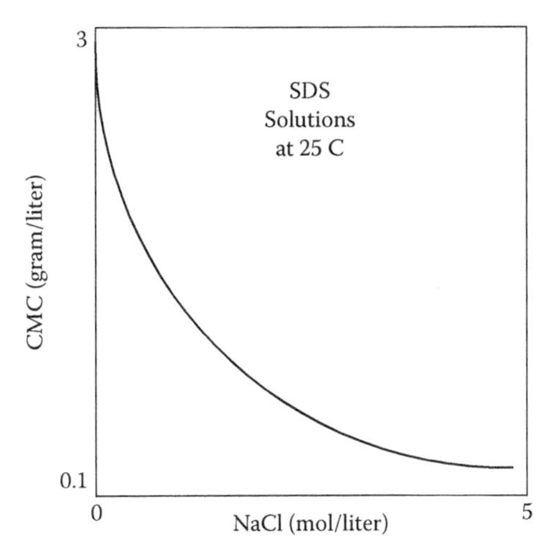

FIGURE 3.12 Variation of CMC with added NaCl for micelles.

is as should be expected from theoretical considerations. It is important to note how the mere addition of 0.01 mole/L of NaCl changes the CMC by 65%. This shows that the charge–charge repulsion is very significant and is reduced appreciably by the addition of counterions (in this case, Na+).

The change of CMC with NaCl for SDS is as follows (at 25°C) (Figure 3.10):

NaCl (mole/L)	CMC(mol/L)	g/Liter	N_{agg}
0	0.008	2.3	80
0.01	0.005	1.5	90
0.03	0.003	0.09	100
0.05	0.0023	0.08	104
0.1	0.0015	0.05	110
0.2	0.001	0.02	120
0.4	0.0006	0.015	125

The radius of the spherical micelle is reported as 20 Å, which increases to 23 Å (for the nonspherical).

Experiments have shown that in most cases, such as for SDS, the initial spherical-shaped micelles may grow under some influence into larger aggregates (disc-like, cylindrical, lamellar, vesicle) (Figure 3.13). The spherical micelle has a radius of 17 Å. The extended length of the SDS molecule is about 17 Å. However, larger micelles (as found in 0.6 mole/liter NaCl solution) have dimensions of 17 Å and 25 Å, radii of an ellipse.

It is important to note how CMC changes with even a very small addition of NaCl. This has many other consequences and will be mentioned later in a different context. It has been found in general that the change in CMC with the addition of ions follows the relation:

$$\text{Ln (CMC)} = \text{Constant}_1 - \text{Constant}_2\ [\text{Ln (CMC} + C_{ion})] \tag{3.8}$$

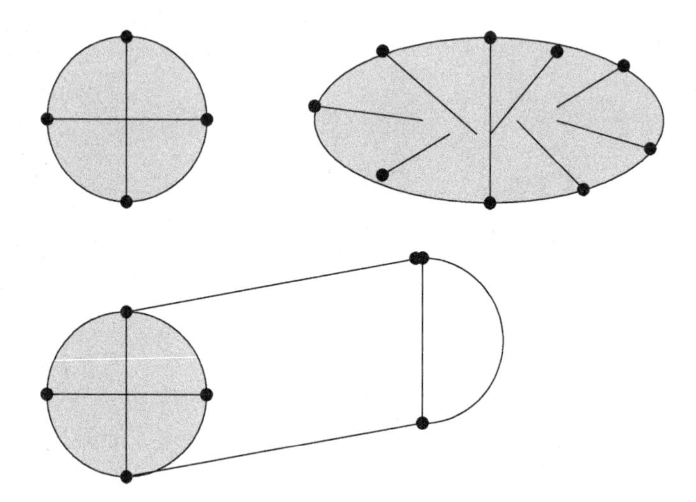

FIGURE 3.13 Different types of micellar aggregates: spherical, disc-like, cylindrical.

The magnitude of $Constant_2$ has been related to the *degree of micelle charge*. Its magnitude has varied from 0.6 to 0.7, which means that micelles have a 30% charge. This means that if there are 100 monomers per micelle, then approximately 70% counterions are bound.

The following data was reported for the CMC of cationic surfactants decreased on the addition of K Br as follows:

DTAB (dodecyltrimethylammonium bromide) and TrTAB:

$$\text{Ln (CMC)} = -6.85 - 0.64 \text{ Ln (CMC} + C_{KBr})$$

TrTAB:

$$\text{Ln (CMC)} = -8.10 - 0.65 \text{ Ln (CMC} + C_{KBr})$$

TTAB:

$$\text{Ln (CMC)} = -9.43 - 0.68 \text{ Ln (CMC} + C_{KBr})$$

It has been noted that the magnitude of slope increases with the increase in alkyl chain length. A similar relationship has been reported for the Na-alkyl sulfate homologous series.

The alkyl chain length has a very significant effect (decrease with increase in n_C) on the CMC. The CMC data for soaps give the following dependence on the alkyl chain length:

Soap	CMC (mole/Liter)25C
C7COOK	0.4
C9COOK	0.1
C11OOK	0.025
C7COOCs	0.4

These data show that CMC decreases by a factor of 4 for each increase in chain length by $-CH_2CH_2-$. This again indicates that due to lower solubility in water with increasing chain length, CMC is related to the latter characteristic of the molecule. Further, this effect will be valid for all kinds of detergent molecules (both with and without charges).

3.3 GIBBS ADSORPTION EQUATION IN SOLUTIONS

A pure liquid (such as water) when shaken does not form any foam. This merely indicates that the surface layer consists of pure liquid (and absence of any minor surface active impurities). However, if a very small amount of surface active agent (soap or detergent, approximately millimole concentration or about parts per million [ppm] by weight) is added, and one then shakes the solution, foam forms at the surface of the solution. This indicates that the surface active agent has accumulated at the surface (meaning that the concentration of the surface active agent is much higher at the surface than in the bulk phase, and in some cases, many thousands of times), and thus forms a thin liquid film that constitutes the bubble. In fact, the bubble or foam formation is useful criteria with regard to the purity of the system. Generally, foam bubbles are formed under different conditions on the shores of lakes or oceans. If water in these sites is polluted, then very stable foams are observed. However, there are also naturally formed surface active substances that contribute to foaming. It must be mentioned that if one adds instead an inorganic salt, NaCl, then no foam is formed. The foam formation indicates that the *surface active agent* adsorbs at the surface, and forms a *thin liquid film* (TLF) (consisting of two layers of amphiphile molecule and with some water). Thin liquid film is

One layer of surface active substance	= = = = = = = = = = = = = = =
Water molecules in between	wwwwwwwwwwwww
One layer of surface active substance	= = = = = = = = = = = = = = =

This may be compared to a sandwich-type of structure. It is important to mention that these bubble structures are of nanometer dimensions but are still easily visible to the naked eye.

This has led to many theoretical analyses of surfactant concentration (in the bulk phase) and the surface tension (which will be related to the presence of surfactant molecules at the surface). The thermodynamics of surface adsorption has been extensively described by Gibbs adsorption theory (Chattoraj and Birdi, 1984). Further, Gibbs adsorption theory is also described for other systems than solutions (such as solid–liquid or $liquid_1$–$liquid_2$; adsorption of solute on polymers). In fact, in any system

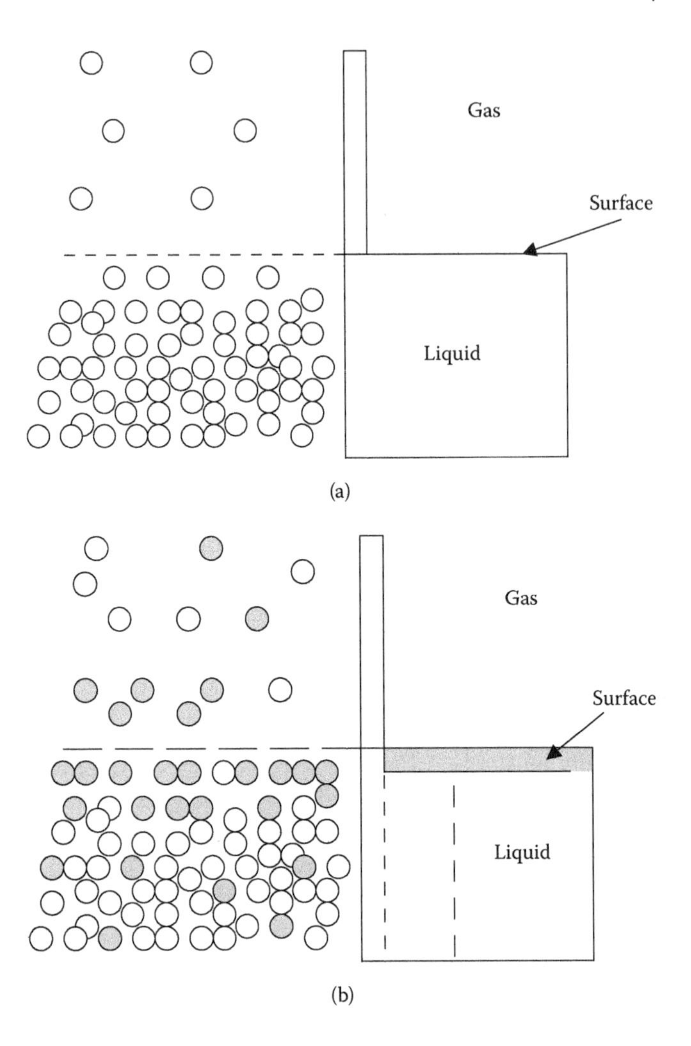

FIGURE 3.14 Surface composition of (a) pure water and (b) an ethanol–water solution (shaded = ethanol).

where adsorption takes place at an interface, the Gibbs theory will be applicable (such as solid–liquid; protein molecule–solution with solutes which may adsorb).

3.3.1 Gibbs Adsorption Theory at Liquid Interfaces

The magnitude of surface tension of water is sensitive to the addition of different molecules. The surface tension of water changes with the addition of organic or inorganic solutes, at constant temperature and pressure (Defay et al., 1966; Chattoraj and Birdi, 1984; Birdi, 1989, 1997). The extent of surface tension change and the sign of change are determined by the molecules involved (see Figure 3.1). The magnitude of surface tension of aqueous solutions generally increases with different electrolyte

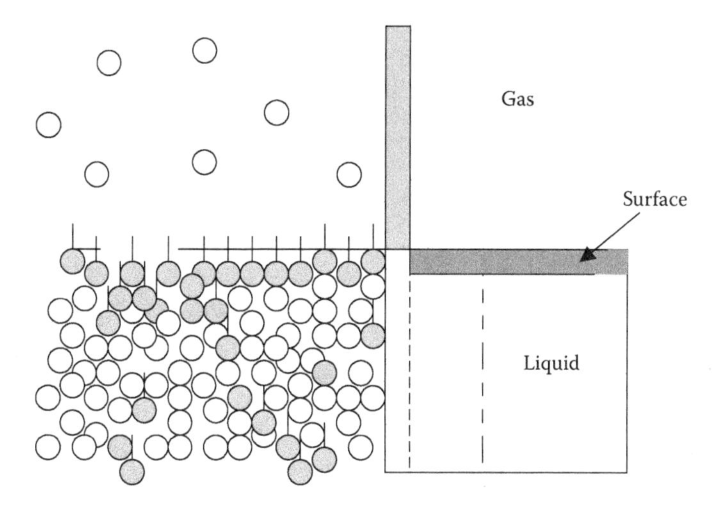

FIGURE 3.15 The concentration of a detergent (shaded with a tail) in a solution and at the surface. The shaded area at the surface is the excess concentration due to accumulation.

concentration. The magnitude of surface tension of aqueous solutions containing organic solutes invariably decrease. As mentioned earlier, the surface of a liquid is where the density of liquid changes to that of a gas, by a factor of 1000 (Chapter 2, Figure 2.2). As an example, let us now look at what happens to surface composition when ethanol is added to water (Figure 3.14).

The reason ethanol concentration in the vapor phase is higher than water is due to its lower boiling point. Next let us consider the situation when a detergent is added to water whereby the surface tension is lowered appreciably (Figure 3.15).

Change in surface tension of water with the addition of different solutes:

Inorganic salts = increase in γ
Organic substances (such as ethanol) = decrease in γ
Soaps or detergents = appreciable decrease in γ

The schematic concentration profile of detergent molecules is such that the concentration is homogeneous up to the surface. At the surface there is almost only detergent molecule plus the necessary number of water molecules (which are in a bound state to the detergent molecule). The surface thus shows very low surface tension, about 30 mN/m. The surface concentration profile of detergent is not easily determined by any direct method. Figure 3.15 is shown as a rectangle for convenience, but one may also imagine other forms of profiles, such as curved. It can be observed that surface tension decreases due to ethanol. This suggests that there are more ethanol molecules at the surface than in the bulk. This is also seen in a cognac glass. The ethanol vapors are observed to condense on the edge of the glass. This shows that the concentration of the surface of the solution is very high. To analyze

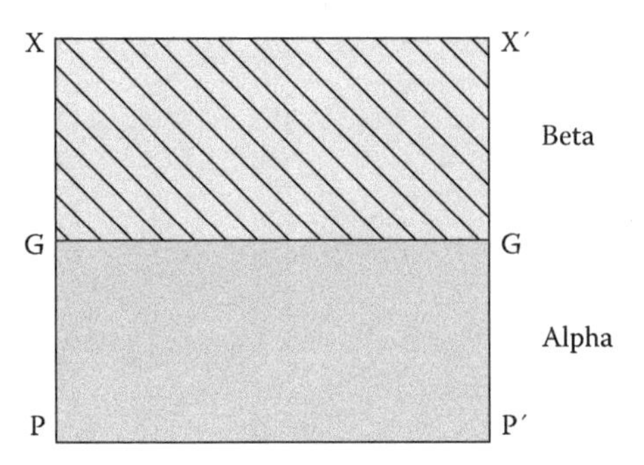

FIGURE 3.16 Liquid column in a real system of α-phase and ß-phase.

these data, the well-known Gibbs adsorption equation needs to be used (Chattoraj and Birdi, 1984; Birdi, 1989; Birdi, 2009).

Derivation of Gibbs Adsorption Equation

The Gibbs adsorption equation has been a subject of many investigations in the literature. Here only a short description is given. The reader may find more detailed description in the references (Chattoraj and Birdi, 1984). A liquid column containing i number of components is shown in Figure 3.16, according to the Gibbs treatment of two bulk phases, that is, α and ß separated by the interfacial region AA′BB′.

Gibbs considered that this interfacial region is inhomogeneous and difficult to define, and he therefore also considered a more simplified case in which the interfacial region is assumed to be a mathematical plane (Figure 3.17).

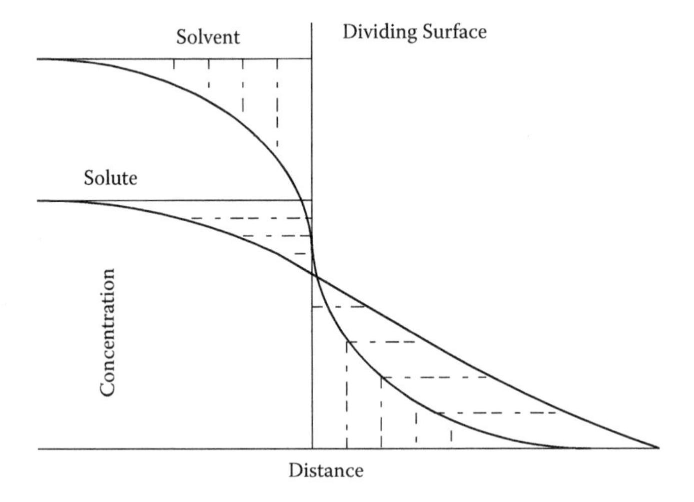

FIGURE 3.17 The liquid composition column in an ideal system.

In the actual system (Figure 3.17) the bulk composition of the i-th component in α and ß phase are $c_{i\alpha}$ and $c_{i\beta}$, respectively. However, in the idealized system, the chemical compositions of the α and ß phases are imagined to remain unchanged right up to the dividing surface so that their concentrations in the two imaginary phases are also $c_{i\alpha}$ and $c_{i\beta}$, respectively. If $n_{i\alpha}$ and $n_{i\beta}$ denote the total moles of the i-th component in the two phases of the idealized system, then the Gibbs surface excess Γ_{ni} of the i-th component can be defined as:

$$n_i^x = n_i^t - n_{i\alpha} - n_{i\beta} \tag{3.9}$$

where n_i^t is the total moles of the i-th component in the real system. In an exactly similar manner, one can define the respective surface excess internal energy, E_x, and entropy, S_x, by the following mathematical relationships (Chattoraj and Birdi, 1984; Birdi, 1989):

$$E^x = E^t - E^\alpha - E^\beta \tag{3.10}$$

$$S^x = S^t - S^\alpha - S^\beta \tag{3.11}$$

Here, E^t and S^t are the total energy and entropy, respectively, of the system as a whole for the actual liquid system. The energy and entropy terms for α and ß phases are denoted by the respective superscripts. The excess (x) quantities thus refer to the surface molecules in adsorbed state.

The real and idealized systems are open so that following equation can be written:

$$dE^t = T\,dS^t - (p\,dV + p'\,dV' - \gamma\,dA) + \mu_1\,dn_{1t} + \mu_2\,dn_{2t} + \cdots + \mu_i\,dn_i^t \tag{3.12}$$

where V^α and V^β are the actual volumes of each bulk phase, and p and p′ are the respective pressures. Since the volume of the interfacial region was considered to be negligible, $V^t = V^\alpha + V^\beta$. Further, if the surface is almost planar, then $p_\alpha = p'_\beta$, and $(p\,dV^\alpha + p_\beta\,dV^\beta) = p\,dV^t$.

The changes in the internal energy for idealized phases α and ß may similarly be expressed as follows:

$$dE^\alpha = TdS^\alpha - p\,dV^\alpha + \mu_i\,dn_i + \cdots + \mu_{i,}\,dn_{i\alpha} \tag{3.13}$$

and

$$dE^\beta = TdS^\beta - p\,dV^\beta + \mu_i\,dn_i + \cdots + \mu_{i,\beta}\,dn_{i,\beta} \tag{3.14}$$

In the real system, the contribution due to the change of the surface energy, γ dA, is included as an additional work. Such a contribution is absent in the idealized system containing only two bulk phases without the existence of any physical interface.

By subtracting Equations (3.26) and (3.27) from Equation (3.25), the following relationship is obtained:

$$d(E^t - E^\alpha - E^\beta) = T\,d(S^t - S^\alpha - S^\beta) + \gamma\,dA + \mu_1\,d(n_t - n_1 - n_{1\alpha})$$
$$+ \cdots + \mu_i\,d(n_{i,t} - n_i - n_{i\beta}) \tag{3.15}$$

or

$$d\,E^x = T\,dS^x + \gamma\,dA + \mu_1\,dn_1\,x + \cdots + \mu_i\,dn_{ix} \quad (3.16)$$

This equation on integration at constant T, γ, and μ_i, gives:

$$E^x = T\,S^x + \gamma\,\mathbf{A} + \mu_1 n_1 ix + \cdots + \mu_1\,n_{ix} \quad (3.17)$$

This relationship may be differentiated in general to give:

$$dE^x = T\,dS^x + \gamma\,dA + 3_i\,(\mu_i\,dn_i) + 3_i n_i d\mu_i + A\,d\gamma + S\,dT \quad (3.18)$$

Combining Equations (3.18) and (3.17) gives:

$$-\mathbf{A}d\gamma = S^x dT + 3_i n_{ix} d\mu_i \quad (3.19)$$

Let $S^{s,x}$ and Γ_{ix} denote the surface excess entropy and moles of the i-th component per surface area, respectively. This gives:

$$S^{s,x} = S^x/A$$

$$\Gamma_{ix} = n_{ix}/A \quad (3.20)$$

and

$$-d\,\gamma = S^{s,x}\,dT + \Gamma_{x,1}\,d\mu_1 + \Gamma_{x,2}\,d\mu_2 + \cdots + \Gamma_{ix}\,d\mu_i \quad (3.21)$$

This equation is similar to the Gibbs-Duhem equations for the bulk liquid system.

To make this relationship more meaningful, Gibbs further pointed out that the position of the plane may be shifted parallel to GG′ along the x direction and fixed in a particular location when (n_{it}) becomes equal to $(n_1 + n_i)$. Under this condition, n_{ix} (n_1 by convention) becomes zero. The relation in Equation (3.21) can be rewritten as:

$$-d\,\gamma = S_{s,1}\,dT + \Gamma_2\,d\mu_2 + \cdots + \Gamma_i\,d\mu_i \quad (3.22)$$

$$= S_{s,1}\,dT + 3\,\Gamma_i\,d\mu_i \quad (3.23)$$

At constant T and p, for a two component system (say, water(1) + alcohol(2)), we thus obtain the classical Gibbs adsorption equation as:

$$\Gamma_2 = -(d\,\gamma/d\mu_2)_{T,p} \quad (3.24)$$

The chemical potential μ_2 is related to the activity of alcohol by the equation:

$$\mu_2 = \mu_2^o + R\,T\ln(a_2) \quad (3.25)$$

If the activity coefficient can be assumed to be equal to unity, then:

$$\mu_2 = \mu^o{}_2 + R\,T \ln (C_2) \tag{3.26}$$

where C_2 is the bulk concentration of solute 2.

The Gibbs adsorption then can be written as:

$$\Gamma_2 = -1/RT(d\gamma/d \ln (C_2))$$

$$= -C_2/RT\ (d\gamma/dC_2) \tag{3.27}$$

This shows that the *surface excess* quantity on the left-hand side is proportional to the change in surface tension with concentration of the solute $(\gamma/d(\ln(C_{surfaceactivesubstance}))$. A plot of ln (C_2) versus γ gives a slope equal to Γ_2 (RT). From this one can estimate the value of Γ_2 (moles/area).

This shows that all surface active substances will always have a higher concentration at the surface than in the bulk of the solution. This relation has been verified by using radioactive tracers. Further, as will be shown later, under spread monolayers there is very convincing support for this relation and the magnitudes of Γ for various systems. The surface tension of water (72 mN/m, at 25°C) decreases to 63 mN/m in a solution of SDS of concentration 1.7 mmol/L. The large decrease in surface tension suggests that SDS molecules are concentrated at the surface, as otherwise there should be very little change in surface tension. This means that the concentration of SDS at the surface is much higher than in the bulk. The molar ratio of SDS:water in the bulk is 0.002:55.5. At the surface the ratio will be expected to be of a completely different value, as found from the value of Γ (ratio is 1000:1). This is also obvious when considering that foam bubbles form on solutions with very low surface active agent concentrations. The foam bubble consists of a bilayer of surface active agent with water inside. It is easy to consider the state of surfactant solutions in terms of molecular ratios.

Example 3.1: Sodium Dodecyl Sulfate (SDS) Aqueous Solution

SDS is a strong electrolyte and in water it dissociates almost completely (at concentrations below CMC):

$$C_{12}\,H_{25}\,SO_4\,Na \equiv C_{12}H_{25}SO_4^- + Na^+$$

$$SDS \equiv DS^- + S^{+\prime} \tag{3.38}$$

The appropriate form of the Gibbs equation will be:

$$-d\,\gamma = \Gamma_{DS^-}\,d\mu_{DS^-} + \Gamma_{S+}\,d\mu_{s+} \tag{3.39}$$

where surface excess, Γ_i, terms for each species in the solution, for example, DS^- and S^+ are included. This equation relates the observed change in st to the changes in the chemical potential of the respective solutes (here DS–, S+) One can expand the chemical potential terms in analogous way to Equation (3.14), then we obtain

$$-d\,\gamma = RT\,[\Gamma_{DS} - d(\ln C_{DS-}) + \Gamma_S + d(\ln C_{s+})] \tag{3.30}$$

$$= RT\,(\Gamma_{DS} - dC_{DS-}/C_{DS-} + \Gamma_S + dCs+/dCs+) \tag{3.30a}$$

In order to simplify this equation we must assume electrical neutrality is maintained in the interface, then we may write:

$$\Gamma_{SDS} = \Gamma_{DS-} = \Gamma_{S+} \tag{3.31}$$

and

$$C_{SDS} = C_{DS-} = C_{S+} \tag{3.32}$$

which on substitution in Equation (3.30) gives:

$$-d\,\gamma = 2\,RT\,\Gamma_{SDS}\,d(\ln C_{SDS}) \tag{3.33}$$

$$= (2RT/C_{SDS})\,\Gamma_{SDS}\,dC_{SDS} \tag{3.34}$$

and one obtains

$$\Gamma_{SDS} = -\tfrac{1}{2}\,(RT)\,d\,\gamma/(d\ln C_{SDS}) \tag{3.35}$$

In the case where the ion strength is kept constant, that is, in the presence of added NaCl, then the equation becomes:

$$\Gamma_{SDS} = -1/(RT)\,[d\,\gamma/(d\ln C_{SDS})] \tag{3.36}$$

Comparing Equation (3.35) with Equation (3.36), it will be seen that they differ by a factor of 2, and that the appropriate form will need to be used in experimental tests of the Gibbs equation. It is also quite clear that any partial of ionization would lead to considerable difficulty in applying the Gibbs equation. Further, if SDS were investigated in a solution using a large excess of sodium ions, produced by the addition of, say, sodium chloride, then the sodium ion term in Equation (3.35) will vanish and will arrive back at an equation equivalent to Equation (3.46).

A quantitative analysis of the adsorption of detergents at the surface of solution can be estimated by the application of the Gibbs equation to the experimental data. The procedure used is to plot γ versus the log ($C_{detergent}$). From γ versus $C_{alkyl\ sulfate}$ data, the following data as obtained from the Gibbs equation is found:

Concentration (mol/L)	$\Gamma_{Salkylsulfate}$ (10^{-12}mol/cm^2)	A(area/molecule)
NaC10 sulfate 0.03 mol/L	3.3	50 Å^2
NaC12 sulfate 0.008mol/L	3.4	50 Å^2
NaC14 sulfate 0.002 mol/L	3.3	50 Å^2

In the plots of γ versus concentration, the slope is related to the surface excess, $\Gamma_{Salkylsulfate}$.

In the absence of any other additives, one gets the following relationship between change in γ and the change in NaSDS concentration (at a given T):

$$\Gamma_{SDS} = -1/(4.606\ R\ T)\ [d\ \gamma/(d\ \log C_{SDS})]$$

The magnitude of Γ thus obtained at the interface thus provides information about the orientation and packing of the NaSDS molecule. This information is otherwise not available by any other means.

The area/molecule values indicate that the molecules are aligned vertically on the surface, irrespective of the alkyl chain length. If the molecules were oriented flat then the value of area/molecule would be much larger (approximately 100 Å^2). Further, since the alkyl chain length has no effect on the area also proves this assumption. These conclusions have been verified from spread monolayer studies. Further, one also finds that the polar group, that is, $-SO^{4-}$, would occupy something like 50 Å^2. Later, it will be shown that other studies confirm that the area per molecule is approximately 50 Å^2.

The Gibbs adsorption equation is a relation about the solvent and a solute (or many solutes). The solute is present either as excess (if there is an excess surface concentration) if the solute decreases the surface tension or a deficient solute concentration (if surface tension is increased by the addition of the solute).

Let us consider the system water to which a surfactant (soap, etc.), such as SDS (SDS dissociates into SD^- and S^+), is added, the molecules at the surface will change as follows:

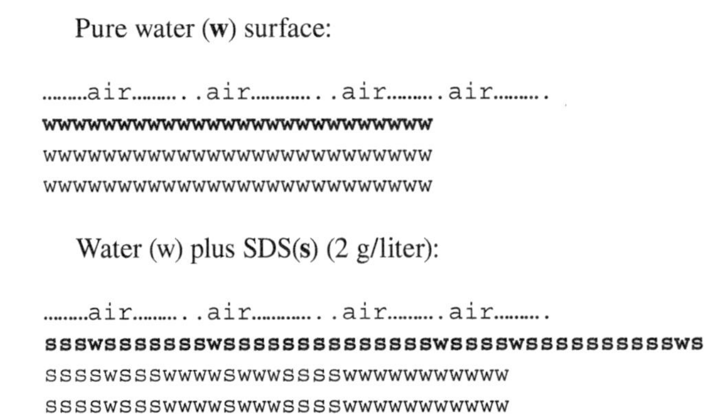

This shows that surface tension of pure water of 72 mN/m decreases to 30 mN/m by the addition of 2 g/liter of SDS. Thus the surface of SDS solution is mostly a monolayer of SDS plus some bound water. The ratio of water:SDS in the system is roughly as follows:

In bulk phase—55 moles water:8 mmole SDS
At the surface—Roughly 100 mole SDS:1 mole water

This description is in accord with the decrease in surface tension of the system.

Investigations have shown that if one carefully sucked a small amount of surface solution of a surfactant, then one can estimate the magnitude of Γ. The concentration of the surface active substance was found to be 8 μmol per mL. The concentration in the bulk phase was 4 μmol/L. The data show that the surface excess is 8 μmol/ml – 4 μmol/ml = 4 μmol/ml. Further, this indicates that when there is 8 μmol/L in the bulk of the solution at the surface the SDS molecules completely cover the surface. The consequence of this is that at higher concentrations than 8 μmol/L no more adsorption at the interface of SDS takes place. Thus, surface tension remains constant (almost). This means that the surface is completely covered with SDS molecules. The area per molecule data (as found to be 50 Å^2) indicates that the SDS molecules are oriented with the SO^{4-} groups pointing toward the water phase while the alkyl chains are oriented away from the water phase. This means if foam bubbles are used, the collected foam would continuously remove more and more surface active substance from the surface. This method of bubble foam separation has been used to purify wastewater of surface active substances (Birdi, 2009). The latter method is especially useful when very minute amounts of surface active substances (dyes in printing industry, pollutants in wastewater) need to be removed. It is economical and free of any chemicals or filters. In fact, if the pollutant is very expensive or poisonous, then this method can have many advantages over the other methods. A simple example is given to understand the useful application of bubbles. Similar estimations can be carried out for other surfactants or other surface active molecules.

Calculation of amount of SDS in each bubble:

Bubble of radius = 1 cm

Assuming that there is almost no water in the bilayer of the bubble (this is a reasonable assumption in the case of very thin films), then the surface area of the bubble can be used to estimate the amount of SDS.

$$\text{Surface area of bubble} = (4\ \Pi\ 1^2)\ 2 = 25\ \text{cm}^2 = 25\ 10^{16}\ \text{Å}^2$$

$$\text{Area per SDS molecule (as found from other studies)} = 50\ \text{Å}^2/\text{molecule SDS}$$

$$\text{Number of SDS molecules per bubble} = 25\ 10^{16}/50 = 0.5\ 10^{16}\ \text{molecules}$$

$$\text{Amount of SDS per bubble} = 0.5\ 10^{16}\ /6\ 10^{23}\ \text{gram}$$

$$= 0.01\ \mu\text{g SDS}$$

These data show that it would require 100 million bubbles to remove 1 g of SDS from the solution! This seems to be a very large number. Since bubbles can be easily produced at very fast rates (about 100 to 1000 bubbles per minute), thus this is not a big hindrance. Consequently, any kind of other surface active substance (such as pollutants in industry) can thus be removed by foaming. However, in general, the concentration of a pollutant substance is of the order of a few mg/liter. For example, in the recycle process in the paper industry, ink pollutant molecules are removed by bubble foam technology. It is also important to consider that if an impurity in water was a surface active molecule, then this procedure can be used to purify water.

During the past decades, a few experiments have been done where verification of Gibbs adsorption has been reported. One of these methods has been carried out by removing by a microtone blade from the thin layer surface of a surfactant solution. Actually, this is almost the same as the procedure for bubble extraction or merely by a careful suction of the surface layer of solution. The surface excess data for a solution of SDS were found to be acceptable. The experimental data was $1.57\ 10^{-18}$ mole cm^{-2}, while from the Gibbs adsorption equation one expected it to be $1.44\ 10^{-18}$ mole cm^{-2}.

Example 3.2

Solution of CTAB shows the following data:

$$\gamma = 47\ \text{mN/m},\ C_{ctab} = 0.6\ \text{mmol/L}$$

$$\gamma = 39\ \text{mN/m},\ C_{ctab} = 0.96\ \text{mmol/L}$$

From these equations one can calculate:

$$d\gamma/d\log(C_{ctab}) = (47 - 39)/(1n(0.6) - 1n(0.96)) = 8/(-0.47) = 17$$

From these data, the area/molecule for CTAB is 90 Å^2, which is reasonable.

The Gibbs adsorption equation thus shows that near the CMC the surfactant molecules are oriented horizontally with the alkyl chains pointing up while the polar groups are interacting with the water. Accordingly, if the data of systems with varying alkyl chain lengths are analyzed, say C_8SO_4Na and $C_{12}SO_4Na$, then the area per molecule should be the same. This is indeed the case according to experimental data.

Kinetic aspects of surface tension of detergent aqueous solutions: It goes without exception, that one needs the information about kinetic aspects of any phenomena. In the present case, one would like to ask how fast the surface tension of a detergent solution reaches equilibrium after it is freshly created.

Let us examine what happens to the surface tension of a detergent solution if a detergent solution is poured into a container. At almost the instantaneous time concentration of the detergent will be uniform throughout the system, that is, it will be the same in the bulk and at the surface. Since the concentration of surface active substance is very low then the surface tension of solution will be the same as of pure water (i.e., 72 mN/m, at 25°C). This is due to the fact that the surface excess at time zero is zero. However, it has been found that the freshly formed surface of a detergent solution exhibits varying rates of change in surface tension with time. A solution is uniform in solute concentration until a surface is created. At the surface, the surface active substance will accumulate dependent on time, and accordingly surface tension will decrease with time. In some cases, the rate of adsorption at the surface is very fast (less than a second), while in other cases it may take longer (Figure 3.18). A freshly created aqueous solution shows surface tension of almost pure water, that is 70 mN/m. However, surface tension starts to decrease rapidly and

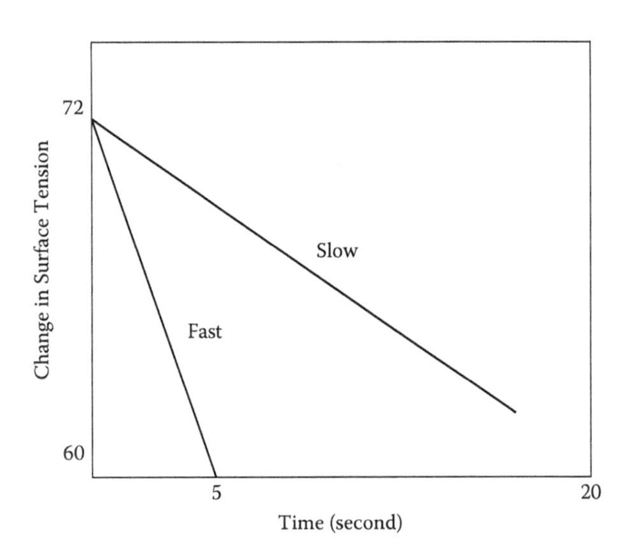

FIGURE 3.18 Kinetics of adsorption of a detergent at an interface.

reaches an equilibrium value (which may be lower than 40 mN/m) after a given time. In general, this phenomena has no consequence. However, in some cases where a fast cleaning process is involved, then one must consider the kinetic aspects. In fact, the formation of foam bubbles as one pours the solution is indicative of that surface adsorption is indeed very fast (as pure water does not foam on shaking!).

Especially, in the case of high molecular weight surface active substances (such as proteins), the period of change in surface tension may be sufficiently prolonged to allow easy observations. This arises from the fact that proteins are surface active. All proteins behave as surface active substances because of the presence of hydrophilic–lipophilic properties (imparted from the different *polar*, such as glutamine and lysine, and *apolar*, such as alanine, valine, phenylalanine, and isovaline, amino acids). Proteins have been extensively investigated with regard to their polar–apolar characteristics as determined by surface activity.

Based on simple diffusion assumptions, the rate of adsorption at the surface

$$d\Gamma/dt = (D/\pi)^2\, C_{bulk}\, t^{-2} \tag{3.37}$$

which on integration gives:

$$\Gamma = 2\, C_{bulk}\, (D\, t/\pi)^2 \tag{3.38}$$

where D is the diffusion constant coefficient and C_{bulk} is the bulk concentration of the solute.

The following procedure has been used to investigate the surface adsorption kinetics.

Solution surface at equilibrium, low γ
After suction at the surface, high γ (almost pure water)
After some time, γ is at equilibrium

In the literature, suction at the surface has been used to investigate the surface adsorption kinetics using high-speed measurement techniques (Birdi, 1989).

The magnitude of surface tension increases right after suction, corresponding to pure water (i.e., 72 mN/m) and decreases with time as Γ increases (from an initial value of zero). This experiment actually verifies the various assumptions as made in the Gibbs adsorption equation. Experimental data shows good correlation to this equation when the magnitude of t is very small. It is also obvious that different detergents will exhibit different adsorption rates. This property will affect the functional properties.

3.3.2 Solubilization in Micellar Solutions of Organic-Water-Insoluble Molecules

In everyday life, systems that involve *organic-water-insoluble compounds* (both in industry and biology) are found. In many of these systems, one is interested in the mechanism of solubility of such organic compounds in water. One of the most important examples is the solubility of a medical compound for pharmaceutical applications. It has been found that micelles (both ionic and nonionic) behave as a microphase, where the *inner core* behaves as (liquid) alkane, while the surface area behaves as a polar phase (Tanford, 1980). The inner core is also found to exhibit liquid-alkane-like characteristics. The inner core thus has been found to exhibit alkane-like properties while being surrounded by a water phase. In fact, micelles are *nanostructures*. What this then suggests is that one can design surfactant solution systems in water, which can have both aqueous and alkane-like properties. This unique property is one of the main applications of surfactant micelle solutions in all kinds of systems (especially in washing and cleaning, cosmetics, pharmaceutical, oil and gas recovery). Further, in ionic surfactant micelles, one can additionally create *nanoreactor* systems. In the latter, reactors of the counterions are designed to bring two reactants to a very close proximity (due to the electrical double layer [EDL]). These reactions would otherwise have been impossible (Birdi, 2007, 2010b).

The most useful characteristic of micelles arises from its inner (alkyl chains) part (Figure 3.19).

Alkyl chains attract each other, thus the inner part consists of alkyl groups that are closely packed. It is known (from experimental data) that these clusters behave as *liquid paraffin* ($C_n H_{2n+2}$). The alkyl chains are thus not fully extended. Hence, this inner hydrophobic part of micelles should exhibit properties that are common for alkanes, such as the ability to solubilize all kinds of water-insoluble organic compounds. The solute enters the alkyl core of the micelle and it swells. The equilibrium is reached when the ratio between moles solute:moles detergent is reached corresponding to the thermodynamic value. Size analyses of micelles (by using light scatter) of some spherical micelles of SDS have indeed shown that the radius of the micelle is almost the same as the length of the SDS molecule. However, if the solute interferes with the outer polar part of the micelle then the micelle system may change, such that the CMC and other properties change. This is observed in the case of dodecanol addition to SDS solutions. However, very small additions of solutes show very little effect on CMC. The data in Figure 3.20 show the change in solubility of naphthalene in SDS aqueous solutions.

Below CMC the amount dissolved remains constant, which corresponds to its solubility in pure water. The slope of the plot above CMC corresponds to 14 mole SDS:1 mole naphthalene. It is seen that at CMC the solubility of naphthalene abruptly increases. This is due to the fact that all micelles can solubilize water-insoluble organic compounds. A more useful analysis can be carried out by considering the thermodynamics of this solubilization process.

At equilibrium, the chemical potential of a solute (naphthalene, etc.) will be given as:

$$\mu_s^{s} = \mu_s^{aq} = \mu_s^{M} \tag{3.39}$$

where μ_s^{s}, μ_s^{aq}, and μ_s^{M} are the chemical potentials of the solute in the solid state, aqueous phase, and micellar phase, respectively. This equilibrium state is common in all kinds of physic/chemical systems. It must be noted that in these micellar solutions we will describe the system in terms of aqueous phase and micellar phase. Before CMC, the solute will only be present in the water phase (as if no detergent

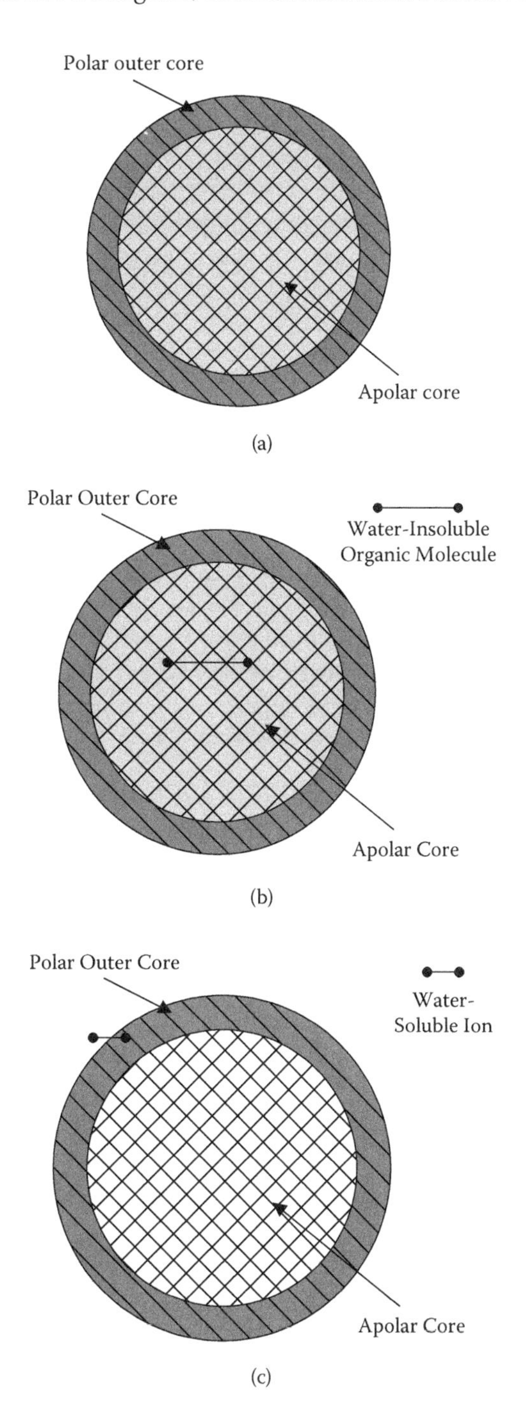

FIGURE 3.19 Micelle structure: (a) inner part = liquid paraffin-like, outer polar part; (b) solubilization of apolar molecule; (c) binding of counterion to the polar part.

is added). The standard free energy change involved in the solubilization, ΔG^{o}_{so}, is given as follows:

$$\Delta G^{o}_{so} = -R\ T \ln (C_{s,M}/C_{s,aq}) \tag{3.40}$$

where $C_{s,aq}$ and $C_{s,M}$ are the concentrations of the solute in the aqueous phase and in the micellar phase, respectively. The free energy change is the difference between the energy when solute is transferred from solid state to the micelle interior. It has been found from many systematic studies that ΔG^{o}_{so} is dependent on the chain length of the alkyl group of the surfactant. The magnitude of ΔG^{o}_{so} changes by –837 J (–200 cal)/mole, with the addition of a -CH2- group. In most cases the addition of electrolytes to the solution has no effect (Birdi, 1982, 1999, 2002). The kinetics of solubilization has an effect on its applications (Birdi, 2002).

Another important aspect is that the slope in Figure 3.20 corresponds to ($1/C_{sM}$). This allows one to determine moles SDS required to solubilize 1 mole of solute. This magnitude is useful in understanding the mechanism of solubilization in micellar systems.

Analyses of various solutes in SDS micellar systems showed that:

Naphthalene, 14 moles SDS/mole naphthalene
Anthracene, 780 moles SDS/mole anthracene
Phenanthrene, 47 moles SDS/mole phenanthrene

The ratio of detergent:solute (in the case of naphthalene) decreases as the chain length of the latter increases. This kind of study thus allows one to determine (quantitatively) the range of solubilization in any such application. These systems when used to solubilize water-insoluble organic compounds would require such information (in such systems as pharmaceutical, agriculture sprays, paints, etc.). Dosage of

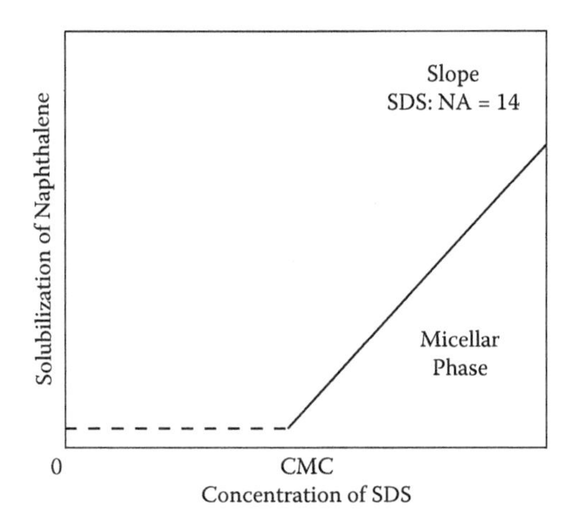

FIGURE 3.20 Solubilization of naphthalene in SDS solutions.

any substance is based upon the amount of material per volume of a solution. This thus also shows that wherever detergents are employed, the major role (in addition to lower surface tension) would be the solubilization of any water-insoluble organic compounds. This process would then assist in the cleaning or washing or any other effect. In some cases, such as bile salts, the solubilization of lipids (especially lecithins) gives rise to some complicated micellar structures. Due to the formation of mixed lipid–bile salt micelles, changes in CMC and aggregation numbers are observed. This has major consequences in the bile salts in biology. Dietary fat consists essentially of mixed triglycerides. These fatty lipids pass through the stomach into the small intestine, without much change in structure. In the small intestine, the triglycerides are partly hydrolyzed by an enzyme (lipase), which leads to the formation of oil–water emulsion. This shows the importance of surface agents in biological systems, such as the stomach.

3.3.3 Biological Micelles (Bile Salt Micelles)

Bile salts are steroids with detergent properties that are used by nature to emulsify lipids in foodstuff passing through the intestine to enable fat digestion and absorption through the intestinal wall. They are secreted from the liver stored in the gallbladder and passed through the bile duct into the intestine when food is passing through. Bile salts in general form micelles with low aggregation numbers (about 10–50) (Tanford, 1980). However, bile micelles grow very large in size when they solubilize lipids. Biosynthesis represents the major metabolic fate of cholesterol, accounting for more than half of the 800 mg/day of cholesterol that the average adult uses up in metabolic processes. By comparison, steroid hormone biosynthesis consumes only about 50 mg of cholesterol per day. Much more than 400 mg of bile salts is required and secreted into the intestine per day, and this is achieved by recycling the bile salts. Most of the bile salts secreted into the upper region of the small intestine are absorbed along with the dietary lipids that they emulsified at the lower end of the small intestine. They are separated from the dietary lipid and returned to the liver for reuse. Recycling thus enables 20 g to 30 g of bile salts to be secreted into the small intestine each day. The most abundant of the bile salts in humans are cholate and deoxycholate, and they are normally conjugated with either glycine or taurine to give glycocholate or taurocholate, respectively. The cholesterol contained in bile will occasionally accrete into lumps in the gallbladder, forming gallstones. In the absence of bile, fats become indigestible and are instead excreted in feces. This causes significant problems in the parts of the intestine, as normally virtually all fats are absorbed in the duodenum and the intestines and bacterial flora are not adapted to processing fats past this point. The role of bile salt micelles in these biological systems is very important. Necessary treatment is basically based on adding extra bile salts to the treatment.

3.4 APPLICATIONS OF SURFACE ACTIVE AGENTS

The area of industry where surface active agents are applied is extensive and is beyond the capacity to be covered completely in this book. However, some main industrial applications will be described in very general terms. The essential description presented here allows one to investigate any other system related to surface or colloid chemistry.

3.4.1 Washing and Laundry (Dry Cleaning)

The most important application of surface and colloid chemistry principles in everyday is in the systems where cleaning and detergency are involved. Cleaning and detergency is one of the most important phenomena for mankind (with regard to health and welfare and technology), and it has been regarded as such for many centuries. For example, the effect of clean wings of airplanes is of utmost concern in flight security. Mankind has been aware of the role of cleanliness on health and disease for many thousands of years. Many critical diseases, such as AIDS or similar infections, are lesser in incidence in those areas of the world where cleanliness is highest. The term *detergency* is used for such processes as *washing clothes* or *dry cleaning* or cleaning. The substances used are designated as detergents (Zoller, 2008). In all these processes the object is to remove dirt from fabrics or solid surfaces (floors or walls or other surfaces of all kinds). The shampoo is used to clean the hair. Hair consists of portentous material and thus requires different kinds of detergents than when washing clothes or a car. Shampoo should not interact strongly with the hair but it should remove dust particles or other material. Another important requirement is that the ingredients in the shampoo should not damage or irritate the eye or the skin with which it may come in contact. In fact, all shampoos are tested for eye irritation and skin irritation before marketing. In some cases, by merely increasing the viscosity one achieves a great deal of protection. For example, a surfactant solution (approximately 20%), alkyl sulfate with two EO (ethylene oxide groups), gives very high viscosity if a small amount of salt is added. The shampoo industry is highly specialized and large industrial state-of-the art research is applied to this product.

In the case of dry cleaning, the aqueous medium is replaced by a nonaqueous medium (such as tetrachloroethylene). Tetrachloroethylene can dissolve grease from materials that should not be treated by water. Since one also has polar dirt (such as sugars, minerals, etc.), a small amount of water needs to be added. This leads to the formation of the so-called *inverted micelles* (Figure 3.21).

In *inverted micelles* polar dirt is solubilized inside the water phase of the inverted micelle.

Detergents used for rug cleaning are tailor made. In the case of the process where detergent solution is applied as highly foaming state, the principle is as follows. As the dirt is removed by the detergent solution and is present in the foam, the latter dries and can be vacuum cleaned. This requires that the detergent is not hygroscopic

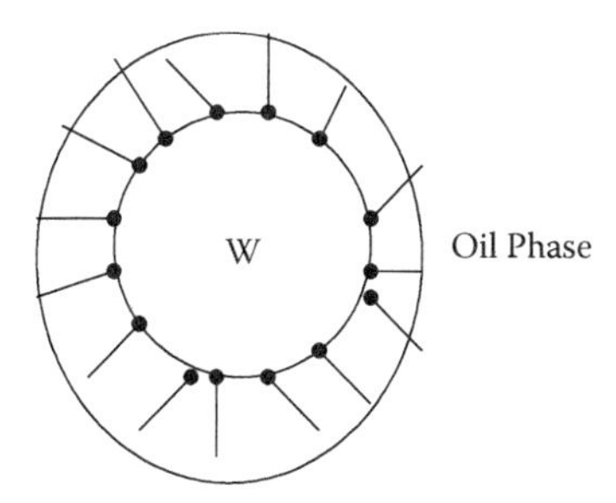

FIGURE 3.21 Structure of inverted micelles. The inner core consists of a water phase.

and thus easily removed. One uses appropriate salts (such as Li+) of alkyl sulfates, which are very weakly hygroscopic. In general, in all such systems, an enhanced detergency when the concentration of detergent is greater than its CMC is observed (Figure 3.22).

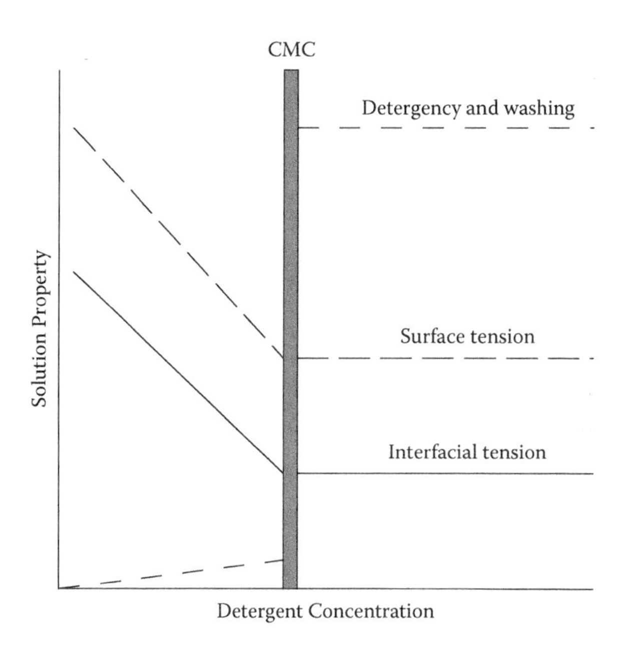

FIGURE 3.22 The change in detergent solution properties as related to CMC.

Appendix 3B: Solubility of Organic Molecules in Water (A Surface Tension–Cavity Model Theory)

Life as we know it on earth is basically dependent on water. However, one also finds that there are a large amount of molecules that are almost insoluble in water. In pure water, all molecules are surrounded by similar molecules in a symmetrical geometry. Of course, this is also the case for any other liquid. In the present we will concern ourselves about what happens when an organic molecule dissolves (to a varying degree) in water. Different kinds of substances dissolve in water, with varying degree. For example, NaCl can dissolve in water up to 10 mole/liter, while an alkane (such as hexane) shows a very low solubility. Let us take a snapshot of what happens when a foreign molecule (organic molecule) is dissolved in water (Figure 3.23).

The process of solubility is as follows. The solute (for example, NaCl) is added to water. NaCl dissociates into Na+ and Cl– ions. These ions are surrounded by water molecules. Ions are known to interact through dipoles and hydrogen bonds with water molecules. On the other hand, when a water-insoluble-organic molecule is placed in water, this system is found to be different than in the case of NaCl or other similar electrolytes. The organic molecule will not interact with dipoles of surrounding water molecules. The solubility of such organic molecules thus requires a cavity in water. It is also essential to visualize that the bigger the cavity needed, the lower is the solubility in water (since cavity requires energy).

Inorganic salts, such as NaCl, dissociate into Na^+ and Cl^- ions and interact with water through hydrogen bonds. Hexane molecule merely dissolves (although very low solubility) in water by placing it inside water structure. Since water structure is stabilized mainly by hydrogen bonds, then hexane molecule will give rise to some rearrangements of these bonds but without breaking.

If a salt exhibits a maximum solubility (called saturation solubility) of 10 mole per liter in water, then it corresponds to about 10 moles of salt:55 moles of water (ratio of 1:5.5). On the other hand, an alkane may show a maximum solubility of 0.0001 mole/liter in water (0.0001 mole:55 mole) or a ratio of 1:550.000. Many decades ago this model was able to predict the solubilities of both simple organic molecules, as

Pure Water

wwwwwwwwwwwwwwwwwwwwwwwwwwwwwwwww

wwwwwwwwwwwwwwwwwwwwwwwwwwwwwwwww

wwwwwwwwwwwwwwwwwwwwwwwwwwwwwwwww

wwwwwwwwwwwwwwwwwwwwwwwwwwwwwwwww

wwwwwwwwwwwwwwwwwwwwwwwwwwwwwwwww

wwwwwwwwwwwwwwwwwwwwwwwwwwwwwwwww

Water with a Solute(S)

wwwwwwwwwwwwwwwwwwwwwwwwwwwwwwwww

wwwwwwwwwwwwwwwwwwwwwwwwwwwwwwwww

wwwssssssssssssssssssssssssssssswwwwwwwwwww

wwwwwwwwwwwwwwwwwwwwwwwwwwwwwwwww

wwwwwwwwwwwwwwwwwwwwwwwwwwwwwwwww

wwwwwwwwwwwwwwwwwwwwwwwwwwwwwwwww

wwwwwwwwwwwwwwwwwwwwwwwwwwwwwwwww

FIGURE 3.23 Pure water (W) and a foreign solute molecule (S) is dissolved in water. The cavity is created for S. (See text for details.)

well as for more complicated cases. In the simplest case, the solubility of heptane is lower than that of hexane, due to the addition of one -CH2- group. In the case of alkane molecules there is found a linear relation between the solubility and the number of -CH2- groups (Tanford, 1980; Birdi, 1997).

This model thus is based on the following assumptions when alkane molecule is placed in water:

- Alkane (CCCCCCCCC) is placed in a cavity in the water (ww).

wwwwwwwwwwwwwwwwwwwwwwwwwwwwwww
wwwwwwwwwwwCCCCCCCCCwwwwwwwwwwww
wwwwwwwwwwwwwwwwwwwwwwwwwwwwwwww

The alkane molecule merely makes a cavity in water without breaking hydrogen bonds of water structure. The energy needed to create a surface area of the cavity will be proportional to the degree of solubility of the alkane. Thus, the solubility of any alkane molecule will be given as:

Free energy of solubility =

$$\text{Proportional to the product (cavity surface area)} \\ \times \text{(Surface tension of the cavity)} \qquad (3B.1)$$

By analyzing the solubility data of a whole range of alkane molecules in water, the following relation was found to fit the experimental data:

Free energy of solubility = ΔG°_{sol}

$$= R\ T\ \mathrm{Log}\ (\mathrm{solubility}) \quad (3B.2)$$

$$= (\gamma_{cavity})\ (S_{areaalkane}) \quad (3B.3)$$

$$= 25.5\ (S_{areaalkane}) \quad (3B.4)$$

For solubility of alkanes in water, the total surface area, TSA, gives the solubility:

$$\ln\ (\mathrm{sol}) = -0.043\ (\mathrm{TSA}) + 11.78 \quad (3B.5)$$

where solubility (sol) is in molar units and TSA in Å^2. For example:

Alkane	(sol)	TSA	Predicted (sol)
N-butane	0.00234	255	0.00143
n-pentane	0.00054	287	0.0004
n-hexane	0.0001	310	0.0001
n-butanol	1.0	272	0.82
N-pentanol	0.26	304	0.21
N-hexanol	0.06	336	0.05

The constant 0.043 is equal to $\gamma_{cavity}/RT = 25.5/600$.

The surface areas of each group in n-nonanol (C9H19OH) were estimated by different methods. These data are as follows:

CH3	CH2	CH2	CH2	CH2	CH2	CH2	CH2	CH2	OH
85	43	32	32	32	32	32	40	45	59

The magnitude of the surface area of the CH3 group is obviously expected to be larger than the CH2 groups. One can estimate the magnitude of TSA of n-decanol, then it will be (TSA of nonanol + TSA of CH2) 431 + 32 = 463 Å^2. From this value one can estimate its solubility.

In oil spill accidents, the solubility of oil in ocean water is of primary concern. This is one of the most important examples where the knowledge of solubility in water is of great importance in science and technology in everyday life.

The data for solubility of homologous series of n-alcohols is of interest, as shown next.

Alcohol	Solubility (mol/L)	Log (S)	Difference per CH2
C4OH	0.97	–0.013	—
C5OH	0.25	–0.60	0.6
C6OH	0.06	–1.22	0.62
C7OH	0.015	–1.83	0.61
C8OH	0.004	–2.42	0.59
C9OH	0.001	–3.01	0.59
C10OH	0.00023	–3.63	0.62

Accordingly, this algorithm allows one to estimate the solubility of water of an organic substance. The estimated solubility of cholesterol was almost in accord with the experimental data (Birdi, 2009). It is seen that log (S) is a linear function of the number of carbon atoms in the alcohol. Each -CH2- group reduces log (S) with 0.06 units.

4 Monomolecular Lipid Films on Liquid Surfaces and Langmuir-Blodgett Films

4.1 INTRODUCTION

Ancient Egyptians are known to have spread very small amounts of oil (olive oil) over water while their ships were sailing into harbors; because small amounts of oil on the water surface were known to appreciably calm the waves thus assisting easy navigation into the harbors. It was later found that some lipid-like substances (almost insoluble in water) formed *self-assembly monolayers* (SAMs) (Figure 4.1) on water surfaces (Gaines, 1966; Chattoraj and Birdi, 1984; Birdi, 1989, 1999, 2002; Adamson and Gast, 1997).

A few decades ago, experiments showed that monomolecular films of lipids could be studied by using rather simple experimental methods (Figure 4.2). It is amazing to find that even a monolayer of lipid film can easily be investigated with very high precision. Furthermore, Langmuir was awarded the Nobel Prize in 1920 for his pioneer monolayer film studies (Gaines, 1966; Birdi, 1999, 2002).

It was already explained earlier that when a surface active agent (such as a surfactant or a soap) is dissolved in water, it adsorbs preferentially at the surface (surface excess, Γ). This means that the concentration of the surface active agent at the surface may be as high as 1000 times more than in the bulk. The decrease in surface tension indicated this and also suggests that only a monolayer is present at the surface. For example, in a solution of SDS of concentration 0.008 mole/Liter, the surface is completely covered with sodium dodecyl sulfate SDS molecules. In this chapter we will consider systems where the system consisting of lipids (almost *insoluble* in water) present as monolayers on water will be analyzed. In these systems almost all the substance applied to the surface (in the range of few micrograms) is supposed to be present at the interface. This means that one knows quantitatively the magnitude of surface concentration (same as the surface excess).

These thin organic films with a thickness of 20 Å (greater than 2 nm) or more are now considered to be very useful structures. Already in 1774, Benjamin Franklin had reported on the effect of very small amounts of oil on the surfaces of water. However, in 1920 Langmuir was awarded the Nobel Prize for studying these monomolecular films in an apparatus as depicted in Figure 4.2.

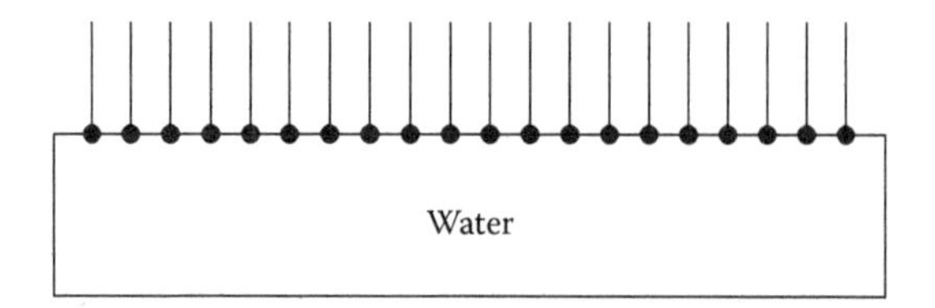

FIGURE 4.1 Lipid monolayer on the surface of water.

If a very small amount of a lipid is placed on the surface of water, it may affect the surface tension in different ways. It may not show any effects (such as cholesterol). It may also show a drastic decrease in surface tension (such as stearic acid or tetradecanol). An amphiphile molecule will adsorb at the air–water or oil–water interface, with its alkyl group pointing away from the water phase. The alkyl group is at a lower energy state when pointing toward air than being surrounded by water molecules. The molecular properties and applications were therefore essential for investigation.

The system can be easily considered as a two-dimensional analog to the classical three-dimensional systems (such as the gas theory).

Later experiments showed that these *monomolecular* films were relevant to many more complicated systems. Thus, it is seen that Π of a monolayer is the lowering of surface tension due to the presence of monomolecular film. This arises from the orientation of the amphiphile molecules at the air–water or oil–water interface, where the polar group would be oriented toward the water phase, whereas the nonpolar part (hydrocarbon) would be oriented away from the aqueous phase. This orientation produces a system with minimum free energy.

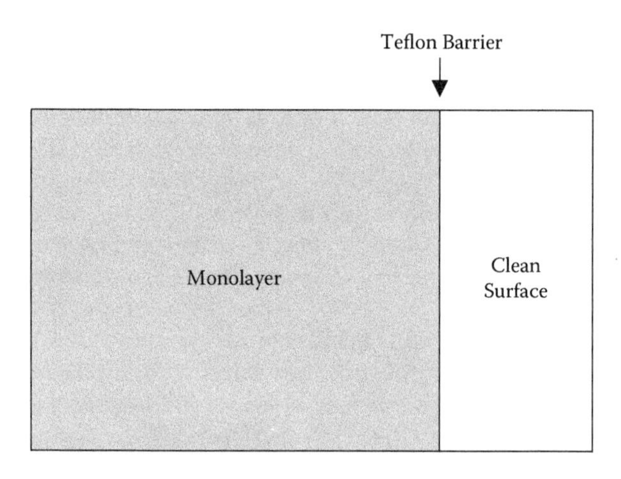

FIGURE.4.2 Monolayer film balance: barrier and lipid film and surface pressure.

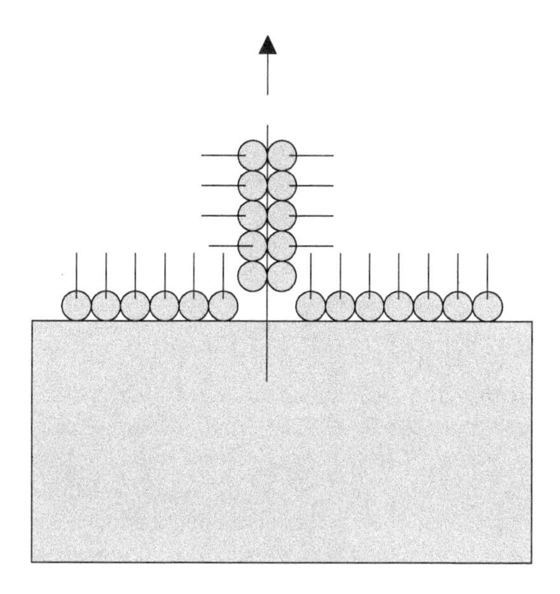

FIGURE 4.3 L-B film formation. (See text for details.)

Later, it was observed that if a clean smooth solid is dipped through monolayer, then in most cases a single layer of lipid would be deposited on the solid. This film was called *Langmuir-Blodgett* (L-B) film (Figure 4.3). Scanning probe microscopes (Birdi, 2003) have been used to verify these structures in very high detail. Further, if this process was repeated, one could deposit multilayers. This L-B film technique has found much application in the electronic industry. Further, the presence of only one layer of lipid changes the surface properties (such as contact angle, wetting, friction, light reflection, charge, adhesion) of the solid.

4.1.1 Apparatus for Surface Film Studies

Since Langmuir reported monolayer studies, a great many instruments (commercial) have been designed about this method. The clean surface of water shows no change in surface tension if a barrier is moved across it. However, if there is a surface active agent present then the latter molecules will be compressed, and this will give rise to a decrease in surface tension.

Monolayer systems are composed of only a monomolecular lipid film on water surface. The aim in these systems is to study the properties of a monomolecular thin lipid film spread on the surface of water, which is not possible by any other technique. Lately, it has been shown that such lipid films are useful membrane models for biological membranes structure and function studies. Modern methods have allowed one to measure the monolayer properties in much detail than earlier. No other method exists through which one can obtain any direct information about the molecular packing or interactions (forces acting in two dimensions).

The monolayer films were studied by using a Teflon trough with a barrier (Teflon) that could move across the surface (Figure 4.2). The change in surface tension (γ)

was monitored by using a Wilhelmy plate method attached to a sensor. The accuracy could be as high as milliN/m.

The film balance (also called a Langmuir trough) consists of a Teflon (or Teflon-lined) rectangular trough (typically 20 cm × 10 cm × 1 cm). Teflon (PTFE, polytetrafluoroethylene) allows one to keep the apparatus clean. Clean water (distilled water and purified with active charcoal to remove any organic minute contaminants) is used to fill the trough just over the edge. A barrier of Teflon is placed on one end of the trough, which is used as a barrier to compress the lipid molecules. Recently, many commercially available film balances are found.

4.1.1.1 Monomolecular Films

Lipid (or protein or other film forming substance) is applied from its solution to the surface. A solution with a concentration of 1 mg/mL is generally used. Lipids are dissolved in $CHCl_3$ or ethanol or hexane (as found suitable). If the surface area of the trough is 100 cm^2, one may use 1 to 100 μL of this solution. After the solvent has evaporated (about 15 minutes) the barrier is made to compress the lipid film at a rate of 1 cm/sec or as suitable. The amount of lipid or protein applied is generally calculated such as to give a compressed film. Most substances cover 1 mg per square meter to give a solid film. If one has 100 cm^2 in the trough, then 10 μg or less of substance is enough for such an experiment. It is thus obvious that such studies can provide much useful surface chemical information with very minute amount of material. Most proteins can also be studied as monomolecular films (1 mg of protein spreads to cover about 1 m^2 surface area) (Birdi, 1989, 1999). This method is useful for studying systems when the magnitude of surface pressure (which is the change in the surface tension) is less than 1 mN/m, and using very high sensitivity apparatus (± 0.001 mN/m).

4.1.1.2 Constant Area Monolayer Film Method

One can also study Π versus surface concentration (C_s = area/molecule) isotherms by keeping the area constant. The surface concentration, C_s, is changed by adding small amounts of a substance to the surface (by using a microliter syringe, 1 or 5 μL). In general, one obtains a good correlation with the surface pressure versus area isotherms (Gaines, 1966; Birdi, 1989, 1999; Adamson and Gast, 1997).

4.1.2 Monolayer Structures on Water Surfaces

Lipids with a suitable hydrophilic–lipophilic balance (HLB) are known to spread on the surface of water to form monolayer films. It is obvious that if the lipid-like molecule is highly soluble in water, then it will disappear into the bulk phase (same as is observed for SDS). Thus, criterion for a monolayer formation is that it exhibits very low solubility in water. The alkyl part of the lipid points away from the water surface. The polar group is attracted to the water molecules and is inside this phase at the surface. This means that the solid crystal when placed on the surface of water is in equilibrium with the film spread on the surface. A detailed analysis of this equilibrium has been given in literature (Gaines, 1966; Adamson and Gast, 1997; Birdi, 1999). The thermodynamics allows one to obtain extensive physical data about this system. It is thus apparent that since one is studying only one monolayer of the substance, the effect of temperature can be very evident.

4.1.3 Self-Assembly Monolayer Formation

The most fascinating characteristic some amphiphile molecules exhibit is that when mixed with water they form *self-assembly structures*. This was already discussed in Chapter 2 on micelle formation. This is in contrast to simple molecules such as methanol or ethanol in water. Since most of the biological lipids also exhibit self-assembly structure formation, this subject has been given much attention in the literature (Birdi, 1989, 1999). The lipid monolayer studies thus provide a very useful method to obtain information about the SAM formation. These studies have provided much information about both technical systems and the cell bilayer structures. Nature has used these lipid molecules throughout the biological world, and the SAM formation has been the basis of all biological cells. A monolayer system thus mimics the cell structure and function.

4.1.4 States of Lipid Monolayers Spread on Water Surfaces

One is quite familiar with the changes in matter going from solid to liquid or vice versa with temperature or pressure. It has been found that even a monolayer of lipid (on water) when compressed can undergo various degrees of packing states. This is somewhat similar to three-dimensional structures (gas–liquid–solid). In the following the various states of monomolecular film will be described as measured from the surface pressure, Π, versus area, A, isotherms, in the case of simple amphiphile molecules. On the other hand, the Π–A isotherms of biopolymers will be described separately, since these are of a different nature.

But before presenting this analysis, it is necessary to consider some parameters of the two-dimensional states that should be of interest. We need to start by considering the physical forces acting between the alkyl–alkyl groups (parts) of the amphiphiles, as well as the interactions between the polar head groups. In the process where two such amphiphiles molecules are brought closer during the Π–A measurement, the interaction forces would undergo certain changes which would be related to the packing of the molecules in the two-dimensional plane at the interface in contact with water (subphase).

This change in packing thus is analogous conceptually to the three-dimensional P-V isotherms, as are well known from the classical physical chemistry (Gaines, 1966; Birdi, 1989; Adamson and Gast, 1997). We know that as pressure, P, is increased on a gas in a container, when $T < T_{cr}$, the molecules approach closer and transition to a liquid phase takes place. Further compression of the *liquid state* results in the formation of a *solid phase*.

In the case of alkanes, the distance between the molecules in the solid state phase is about 5 Å, while it is 5 to 6 Å in the case of liquid state. The distance between molecules in the gas phase, in general, are about $1000^{1/3} = 10$ times larger than in the liquid state (water:volume of 1 mole water = 18 cc; volume of 1 mol gas = 22.4 L). In fact, monomolecular film studies are the only direct method of obtaining such information at interfaces of lipids. Considering that only microgram quantities are enough for such information, the importance of such studies becomes clearly evident. It has been found that the isotherms of the two-dimensional films also resemble three-dimensional P-V isotherms, and the same classical molecular description can be used with regard to the qualitative analyses of the various states.

In the three-dimensional structural buildup, the molecules are in contact with near neighbors, as well as in contact with molecules that may be 5 to 10 molecular dimensions apart (as found from X-ray diffraction). This is apparent from the fact that in liquids there is a long range order up to 5 to 10 molecular dimensions. On the other hand, in the two-dimensional films the state is much different. The amphiphile molecules are oriented at the interface such that the polar groups are pointed toward water (subphase), while the alkyl groups are oriented away from the subphase. This orientation gives the minimum surface energy. The structure is stabilized through lateral interaction between:

1. Alkyl—alkyl groups Attraction
2. Polar group—subphase Attraction
3. Polar group—polar group Repulsion

The alkyl–alkyl groups' attraction arises from the van der Waals forces. The magnitude of van der Waals forces increases with:

- Increase in alkyl chain length
- Decrease in distance between molecules (or when area/molecule [A] decreases).

Experiments show that as the alkyl chain length increases, the magnitude of the surface pressure of the films becomes more stable and gives a high collapse pressure, Π_{co}. The stable films are formed when the attraction forces are stronger than the repulsive forces.

4.1.4.1 Self-Assembly Monolayers

In nature, molecules exhibit very specific characteristics. Man has therefore investigated these molecules exhaustively in order to understand the nature around us. The most important feature one has observed in surface chemistry is the way the surface active substances assemble in monolayers or multilayers. Especially, some lipids (such as lecithins) when dispersed in water form very well-defined assemblies, where the alkyl parts of a molecule are in close proximity with each other. This leads to self-assembly formation with many important consequences. Additionally, micelle formation is one of the most common SAM structures. Further, the essential basis of biological cell structure and function is dependent on the lipid–bilayer structure, which is the SAM phenomenon. Some investigators have even suggested that during the evolution, life was first conceived by some accidental SAM process.

The most convincing results were those as obtained with the normal fatty alcohols and acids. Their monomolecular films were stable, and exhibit very high surface pressures (Birdi, 1989, 1999). A steep rise in surface pressure is observed around 20.5 Å, regardless of the number of carbon atoms in the chains. The volume of a

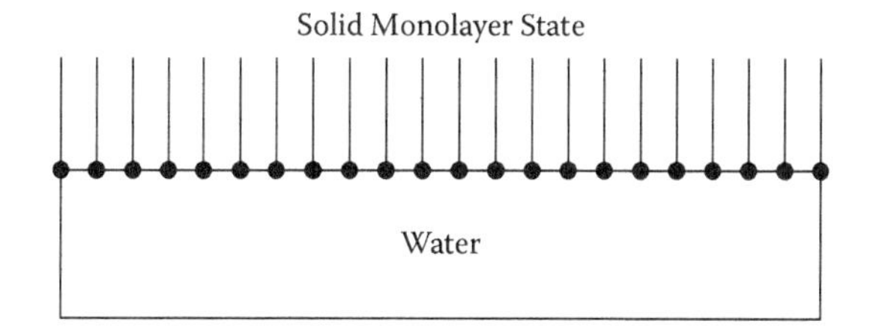

FIGURE 4.4 Orientation of solid films of aliphatic alcohols or acids as films on the surface of water.

-CH2- group is 29.4 Å^3. This gives the length of each -CH2- group perpendicular to the surface, or the vertical height of each group as about 1.42 Å. This compares very satisfactorily with the x-ray data with this value of 1.5 Å. This means that such straight chain lipids are oriented in this compressed state in vertical orientation (Figure 4.4).

As high pressures lead to transitions from gas to liquid to solid phases in the three-dimensional systems, a similar state of affairs would be expected in the two-dimensional film compression surface pressure versus A isotherms (Figure 4.5).

4.1.4.2 Gaseous Monolayer Films

The simplest type of amphiphile monolayer film or a polymer film would be a "gaseous" state. This film would consist of molecules that are at a sufficient distance apart from each other such that lateral adhesion (van der Waals forces) are negligible. However, there is sufficient interaction between the polar group and the subphase that the film forming molecules cannot be easily lost into the gas phase, and that the amphiphiles are almost insoluble in water (subphase).

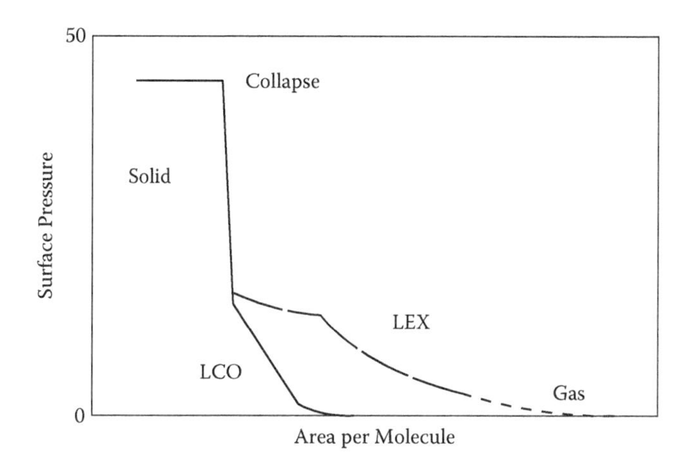

FIGURE 4.5 Lipid monolayer phases (monolayer phases are dependent on both molecular structure and temperature).

When the area available for each molecule is many times larger than molecular dimension, the gaseous-type film (state 1) would be present. As the area available per molecule is reduced, the other states, that is, liquid-expanded (L_{ex}), liquid-condensed (L_{co}), and finally the solid-like (S or solid-condensed) states, would be present.

The molecules will have an average kinetic energy, that is, 1/2 k_BT, for each degree of freedom, where k_B is the Boltzmann constant (= 1.372 10^{-16} ergs/T), and T is the temperature. The surface pressure measured would thus be equal to the collisions between the amphiphiles and the float from the two degrees of freedom of the translational kinetic energy in the two dimensions. It is thus seen that the ideal gas film obeys the relation:

$$\Pi \, A = k_B T \text{ (ideal film)} \tag{4.1}$$

$$\Pi \text{ (mN/m) } A \text{ (Å}^2 \text{ per molecule)} = 411 \text{ (T = 298 K) (ideal film)} \tag{4.2}$$

In Figure 4.6, the Π–A versus A is plotted for an ideal monolayer film. Various molecules have been found to give such ideal films (such as C14H29OH, C15OH, valinomycin, proteins).

This is analogous to the three-dimensional gas law (i.e., PV = k T) (where P is the pressure and V is the volume of the system). At 25°C, the magnitude of (k_BT) = 411 10^{16} ergs. If surface pressure is in millinewton per meter (mN/m) and A in angstroms (Å), then the magnitude of k_BT = 411. In other words, if one has a system with A = 400 Å per molecule, then the value of Π = 1 mN/m is for the ideal gas film.

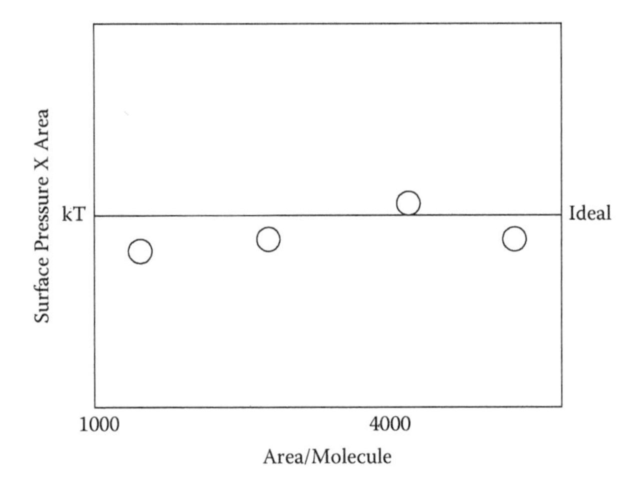

FIGURE 4.6 Π–A versus A plot for an ideal monolayer film: ideal film (Π **A** = 411 = k_B T); C15OH (circles data). (Π in mN/m; A in Å^2/molecule.)

In general, *ideal gas* behavior is only observed when the distances between the amphiphiles are very large, and thus the value of the surface pressure is very small, that is, <0.1 mN/m. It should also be noted that from such sensitive data the molecular weight of the molecule in the monolayer can be estimated. This has been extensively reported for protein monolayers (Birdi, 1989, 1999; Adamson and Gast, 1997). The latter observation requires an instrument with very high sensitivity, ±0.001 mN/m. Modern instruments are easily capable of achieving this sensitivity. The Π versus A isotherms of n-tetradecanol, pentadecanol, pentadecyclic acid, and palmitic acid in the low surface pressure region showed data that agreed with the ideal film. Similar data for isotherms were reported for other lipid monolayers by other workers. The various forces that are known to stabilize the monolayers are mentioned as:

$$\Pi = \Pi_{kin} + \Pi_{vdw} + \Pi_{electro} \tag{4.3}$$

where

Π_{kin} arises from kinetic forces

Π_{vdw} is related to the van der Waals forces acting between the alkyl chains (or groups)

$\Pi_{electro}$ is related to polar group interactions (polar group–water interaction, polar group–polar group repulsion, charge–charge repulsion).

When the magnitude of A is very large, then the distance between molecules is large. If there are no van der Waals or electrostatic interactions, then the film obeys the ideal equation. As the area per molecule is decreased then the other interactions become significant. The Π versus A isotherm can be used to estimate these different interaction forces. These analytical procedures have been extensively described in the current literature (Gaines, 1966; Birdi, 1989, 1999; Adamson and Gast, 1997).

The ideal equation has been modified to fit Π versus A data, in those films where co-area, A_o, correction is needed (Birdi, 1989):

$$\Pi (A - A_o) = k_B T \tag{4.4}$$

In the case of straight chain alcohols or fatty acids, A_o is almost 20 Å^2, which is the same as found from the x-ray diffraction data of the packing area per molecule of solid alkanes.

This equation is thus valid when $A >> A_o$. The magnitude of Π is 0.2 mN/m for A = 2000 Å^2, for ideal film. However, Π will be about 0.2 mN/m for A = 20 Å^2 for a solid-like film of a straight chain alcohol.

Π versus A for a monolayer of valinomycin (a dodeca-cyclic peptide) shows that the relation as given in Equation (4.1) is valid. In this equation it is assumed that the amphiphiles are present as monomers. However, if any association takes place, then the measured values of (Π A) would be less than $k_B\mathbf{T} < 411$, as has also been found (Birdi, 1989, 1999). The magnitude of $k_B T = 4\ 10^{-21}$ J, at 25°C.

In the case of nonideal films, one will find that the versus data does not fit the relation in the equation. This deviation requires that other modified equations of state be used. This procedure is the same as the one used in the case of three-dimensional gas systems.

4.1.4.3 Liquid Expanded and Condensed Films

The Π versus A data is found to provide much detailed information about the state of monolayers at the liquid surface. In Figure 4.6 some typical states are shown. The different states are very extensively analyzed and will be therefore described below.

In the case of simple amphiphiles (fatty acids, fatty alcohols, lecithins, etc.), in several cases, transition phenomena have been observed between the gaseous and the coherent states of films, which show a very striking resemblance to the condensation of vapors to liquids in the three-dimensional systems. The liquid films show various states in the case of some amphiphiles, as shown in Figure 4.6. In fact, if the Π versus **A** data deviates from the ideal equation, then one may expect the following interactions in the film:

- Strong van der Waals
- Charge–charge repulsions
- Strong hydrogen bonding with subphase water

This means that such deviations thus allow one to estimate these interactions.

In general, there are two distinguishable types of liquid films. The first state is called the liquid expanded (L_{ex}) (Gaines, 1966; Chattoraj and Birdi, 1984; Adamson and Gast, 1997). If one extrapolates the Π–A isotherm to zero Π, the value of A obtained is much larger than that obtained for close packed films. This shows that the distance between the molecules is much larger than one will find in the solid film, as will be discussed later. These films exhibit very characteristic elasticity, which will be described further below.

As the area per molecule (or the distance between molecules) is further decreased, there is observed a transition to a so-called liquid condensed (L_{co}) state. These states have also been called "solid expanded" films (Adam, 1941; Gaines, 1968; Birdi, 1989, 1999; Adamson and Gast, 1997), which will be discussed later in further detail. The Π versus A isotherms of n-pentadecylic acid (amphiphile with a single alkyl chain) have been studied, as a function of temperature. Π-A isotherms for two-chain alkyl groups, such as lecithins, also showed a similar behavior.

4.1.4.4 Solid Films

As the film is compressed then a transition to a solid film is observed, which collapses at higher surface pressure. The Π versus A isotherms, below the transition temperatures show the liquid to solid phase transition. These solid films have been also called *condensed films*. These films are observed in such systems where the molecules adhere to each other through the van der Waals forces, very strongly. The Π–**A** isotherm shows generally no change in Π at high A, while at a rather low A value, one observes a sudden increase in Π, as shown in Figure 4.7. In the case of straight chain molecules, such as stearyl alcohol, the sudden increase in Π is found to take place at A = 20 to 22Å^2, at room temperature (that is much lower than the phase transition temperature, to be described later). These analyses have shown that the films may under given experimental conditions show three first-order transition states, that is, (1) transition from the gaseous film to the liquid expanded, L_{ex}; (2) transition from the liquid expanded (L_{ex}) to the liquid condensed (L_{co}); and (3) from L_{ex} or L_{co} to the solid state, if the temperature is below the transition temperature. The temperature above, where no expanded state is observed has been related to the melting point of the lipid monolayer.

4.1.4.5 Collapse States of Monolayer Assemblies

The measurements of Π versus **A** isotherms generally exhibit when compressed a sharp break in the isotherms that has been connected to the *collapse* of the monolayer under the given experimental conditions. The monolayer of some lipids, such as cholesterol, is found to exhibit an usual isotherm (Figure 4.7). The magnitude of surface pressure increases very little as compression takes place. In fact, the collapse state or point is the most useful molecular information from such studies. It will be shown later that this is the only method that can provide information about the structure and orientation of the amphiphile molecule at the surface of water.

However, a steep rise in surface pressure is observed and a distinct break in the isotherm is found at the collapse. This occurs at $\Pi = 40$ mN/m and $A = 40$ Å².

This value of A_{co} corresponds to the cholesterol molecule oriented with the hydroxyl group pointing toward the water phase. Atomic force microscope (AFM) studies of cholesterol as L-B films has shown that domain structures exist (see Chapter 10). This has been found for different collapse lipid monolayers (Birdi, 2003). Different data have provided much information about the orientation of lipid on water (Table 4.1).

It should be mentioned that monolayer studies are the only procedure that allows one to estimate the area per molecule of any molecule as situated at the water surface. In

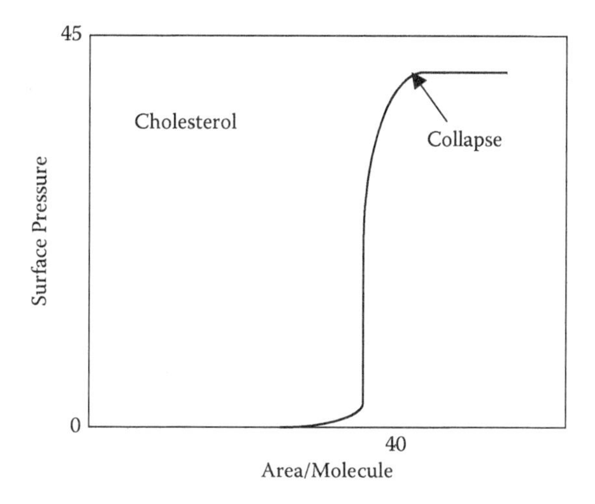

FIGURE 4.7 Π versus A isotherm of cholesterol monolayer on water.

TABLE 4.1
Magnitudes of A_0 for Different Film-Forming Molecules on the Surface of Water

Compound	A_0 (Å^2)
Straight chain acid	20.5
Straight chain acid on dilute HCl	25.1
N-fatty alcohols	21.6
Cholesterol	40
Lecithins	ca.50
Proteins	About 1 m^2/mg
Diverse synthetic polymers (polyamino acids, etc.)	About 1 m^2/mg

Source: Birdi, K. S., *Lipid and Biopolymer Monolayers at Liquid Interfaces*, Plenum Press, New York, 1989.

general, the collapse pressure, Π_{col}, is the highest surface pressure to which a monolayer can be compressed without a detectable movement of the molecules in the films to form a new phase (see Figure 4.7). In other words, this surface pressure will be related to the nature of the substance and the interaction between the subphase and the polar part of the lipid or the polymer molecule. In the literature, one finds that the monolayer of a lipid can be formed by different spreading methods. The monolayer collapse pressure has been shown to provide much information also in the case of protein monolayers.

4.1.5 Diverse Characteristics of Lipid Films

Any reaction between two or more molecules will depend on the geometrical orientation, in addition to other factors. The lipid monolayer is a well-defined structure with the polar part pointing toward the water phase, and the apolar (hydrophobic) part pointing away from the water phase. It is observed that monolayers are very sensitive model to investigate interactions between lipids and other substances (such as ions in water subphase). The long chain fatty acids exhibit gas–L_{ex}–solid film states at high pH. However, the same films on substrates with Ca^{++} or Mg^{++} ions exhibit only solid state. It is also of interest to use this method to investigate the kinetics of penetration by measuring the change in surface pressure with time.

The structure of liquid surfaces with monomolecular films can be studied by measuring the light reflected from the surface. The range of thickness that one generally considers to be measured varies from 100 to 1000 Å (10 to 100 nm). However, in monolayers in which the molecules are oriented and the thickness involved is 5 to 50 Å, the methods have not been easily pursued. In a differential method in which two beams of light from the same incandescent lamp were directed to two similar troughs, reflected light from each was detected by separate photocells. The photocells are arranged in a sensitive bridge circuit so that very small differences can be determined.

The technique of fluorescence spectral measurements has become very sensitive over the past decade. In order to obtain more information on the surface monolayers, a new method based on fluorescence was developed. It consisted of placing the monolayer trough on the stage of an epifluorescence microscope, with doped low concentration of fluorescent lipid probe. Later, ordered solid–liquid coexistence at the water–air interface and on solid substrates was reported. The theory of *domain shapes* has been extensively described by this method.

4.2 OTHER CHANGES AT WATER SURFACES DUE TO LIPID MONOLAYERS

The presence of lipid (or similar kind of substance) monolayer at the surface of aqueous phase gives rise to many changes in the properties at the interface. The major effects, which have been investigated extensively are:

- Surface viscosity (η_s)
- Surface potential (V)

A monomolecular film is resistant to shear stress in the plane of the surface, as also is the case for the bulk phase; a liquid is retarded in its flow by viscous forces. The viscosity of the monolayer may indeed be measured in two dimensions by flow through a canal on a surface or by its drag on a ring in the surface, corresponding to the Ostwald and Couttte instruments for the study of bulk viscosities. The surface viscosity, η_s, is defined by the relation:

$$\text{Tangential force per centimeter of surface} = \eta_s \times \text{Rate of strain} \tag{4.5}$$

and is thus expressed in units of surface poises (m t^{-1}), whereas bulk viscosity, η, is in units of poise (m^{-1} 1 t^{-1}). The relationship between these two kinds of viscosity is:

$$\eta_s = \eta/d \tag{4.6}$$

where d is the thickness of the surface phase, approximately 10^{-7} cm (= 10 Å = 10^{-9} m = nm) for many films. That the magnitude of η_s is of the order of 0.001 to 1 surface poise implies that over the thickness of the monolayer, the surface viscosity is about 10^4 to 10^7 poises. This has been compared to the viscosity comparable to that of butter. The η_s is generally given in surface poise (g/s or kg/s).

Clean water surface consists of only water (W) molecules:

WWWWWWWWWWWWWWWWWWWWWWWWWWWWW Surface

While lipid monolayer (L) gives a surface that is different:

LLLLLLLLLLLLLLLLLLLLLLLLLLLLLLLLLLLLLLLLLLLLL Surface
WWWWWWWWWWWWWWWWWWWWWWWWWWWW First layer
WWWWWWWWWWWWWWWWWWWWWWWWWWWW

Thus, the lipid monolayer gives rise to hindrance to any flow movement at the surface (the layers with *LWLW*). This arises mainly from the fact that water molecules (first layer) bound to the polar part of the lipid are subjected to high viscosity (surface viscosity).

It is easily realized that if a monolayer is moving along the surface under the influence of a gradient of surface pressure, it will carry some of the underlying water with it. In other words, there is no slippage between the monolayer molecules and the adjacent water molecules. The thickness of such regions has been reported to be of the order of 0.003 cm. It has also been asserted that the thickness would be expected to increase as the magnitude of η_s increases. However, analogous to bulk phase, the concept of free volume of fluids should be also considered in these films. As mentioned earlier, the soap films are made of these films when air bubbles form at the liquid surface. Therefore, the soap bubble characteristics are dependent on surface viscosity.

Monolayers of long-chain alcohols exhibit η_s approximately 20 times larger than those of the corresponding fatty acids. This difference may explain some data reported on the effect of temperature on monolayers of these molecules (Birdi, 1989). If the monolayer is flowing along the surface under the influence of π, it carries with it some underlying water. This transport is a consequence of the lack of slippage between the monolayer and the bulk liquid adjacent to it. For a monolayer of oleic acid moving at between 1 and 5 cm/sec, the direct measurement gives the thickness of the entrained water as approximately 0.003 cm. If the bulk viscosity increases, the thickness of the aqueous layer also increases in direct proportion. Accordingly, as mentioned elsewhere, the role of interfacial water needs to be considered in all surface phenomena. It is also obvious that many such films will exhibit complex viscoelastic behavior, the same as those found in bulk phases. The flow behavior can then be treated in terms of viscous and elastic components. Further, the equilibrium elasticity of a monolayer film is related to the compressibility of the monolayer (analogous to the bulk compressibility) by:

$$C^s = -1/A \ (d\ A/d\ \Pi) \tag{4.7}$$

where A is the area per molecule of the film. The surface compressional modulus, K_s (= $1/C^s$), is the reciprocal of C^s. Since there is no change in surface tension with a change in the rate of a pure liquid surface (i.e., $dA/d\ \Pi$ = infinity), the elasticity is thus zero. The interfacial dilational viscosity, k_s, is defined as:

$$\Delta\gamma = k_s \ (1/A)\ (dA/dt) \tag{4.8}$$

where k_s is the fractional change in area per unit time per unit surface tension difference. From this relation C^s and E_s can be derived. Monolayers are thus very useful in understanding various aspects of molecular packing (such as liquid crystals). The information from area/molecule can help estimate the packing and other interaction parameters. These monolayer studies are important for understanding thin liquid film (TLF) structures (e.g., bubbles, foams).

Any liquid surface, especially aqueous solutions, will exhibit asymmetric dipole or ions distribution at surface as compared to the bulk phase. If SDS is present in the bulk solution, then we will expect that the surface will be covered with SD^- ions. This would impart a negative surface charge (as is also found from experiments).

Due to the surface adsorption of SDS to water, not only does surface tension change (reduce) but SDS also imparts negative surface potential (due to SD^- at the surface). Similarly, solutions of cationic detergent, such as CTAB, give rise to a positively charged potential of water surface.

Of course, the surface molecules of methane (in liquid state) obviously will exhibit symmetry in comparison to water molecules. This characteristic can also be associated to the force field resulting from induced dipoles of the adsorbed molecules or spread lipid films (Birdi, 1989; Adamson and Gast, 1997). The surface potential arises from the fact that the lipid molecule orients with the polar part toward the aqueous phase. This gives rise to a change in dipole at the surface. Thus, there would be a change in surface potential when a monolayer is present, as compared to the clean surface. The surface potential, ΔV, is thus:

$$\Delta \mathbf{V} = \text{Potential}_{\text{monolayer}} - \text{Potential}_{\text{clean surface}} = V_{\text{monolayer}} - V_{\text{clean surface}} \qquad (4.9)$$

The magnitude of ΔV is measured most conveniently by placing an air electrode (a radiation emitter, for example, Po^{210} [alpha-emitter]) near the surface (about mm in air) connected to a very high impedance electrometer. This is required since the resistance in air is very high, but it is appreciably reduced by the radiation electrode.

Air electrode— — — — — — -
Monolayer
Water phase electrode— — — — -

The surface voltage is measured as a change in V for pure water surface to that of lipid monolayer.

Since these monolayers are very useful biological cell membrane structures, it is thus seen that such studies can provide information on many systems where ions are actively carried through cell membranes (Chattoraj and Birdi, 1984; Birdi, 1989, 1999).

The transport of K^+ ions through cell membranes by antibiotics (valinomycin) has been a very important example. Addition of K^+ ions to the subphase of a valinomycin monolayer showed that the surface potential became positive. This clearly indicated the ion-specific binding of K^+ to valinomycin (Birdi, 1989). In biological cells the concentrations of different ions is sometimes 20 times different than in the solution outside the cell. The membrane peptides, such as valinomycin, create a channel in the cell membrane for K^+ ions. This leads to the effect that concentration of K^+ ions becomes the same both inside and outside the cell. This leads to the collapse of the cell. Further, the theoretical basis of charged lipid monolayers is well verified from such model monolayer studies.

4.2.1 Charged Lipid Monolayers on Liquid Surfaces

Molecules with charges behave differently than those with no charges. In some instances, there is found that negatively charged molecules behave much differently than positively charged molecules. There is no direct method that can be used to measure these properties. The spread monolayers have provided much useful information about the role of charges at interfaces. In the case of a aqueous solution consisting of fatty acid or SDS, R-Na, and NaCl, for example, the Gibbs equation may be written as (Chattoraj and Birdi, 1984; Birdi, 1989; Adamson and Gast, 1997):

$$-d\gamma = \Gamma_{RNa}\, d\mu_{RNa} + \Gamma_{NaCl}\, d\mu_{NaCl} \tag{4.10}$$

Further:

$$\mu_{RNa} = \mu_R + \mu_{Na} \tag{4.11}$$

$$\mu_{NaCl} = \mu_{Na} + \mu_{Cl} \tag{4.12}$$

It can be easily seen that the following will be valid:

$$\Gamma_{NaCl} = \Gamma_{Cl} \tag{4.13}$$

and

$$\Gamma_{RNa} = \Gamma_R \tag{4.14}$$

It is also seen that following equation will be valid for this system:

$$-d\gamma = \Gamma_{RNa}\, d\mu_{R\,Na} + \Gamma\, d\mu_{Na} + \Gamma\, d\mu_{Cl} \tag{4.15}$$

This is the form of the Gibbs equation for an aqueous solution containing three different ionic species (e.g., R, Na, Cl). Thus, the more general form would be for solutions containing i-number of ionic species as:

$$-d\gamma = \Sigma\, \Gamma_i\, d\mu_i \tag{4.16}$$

In the case of charged film, the interface will acquire a *surface charge*. The surface charge may be positive or negative depending upon the cationic or anionic nature of the lipid or polymer ions. This would lead to the corresponding *surface potential*, φ, also having a positive or negative charge (Chattoraj and Birdi, 1984; Birdi, 1989). The interfacial phase must be electroneutral. This can only be possible if the inorganic counterions also are preferentially adsorbed in the interfacial phase (Figure 4.8).

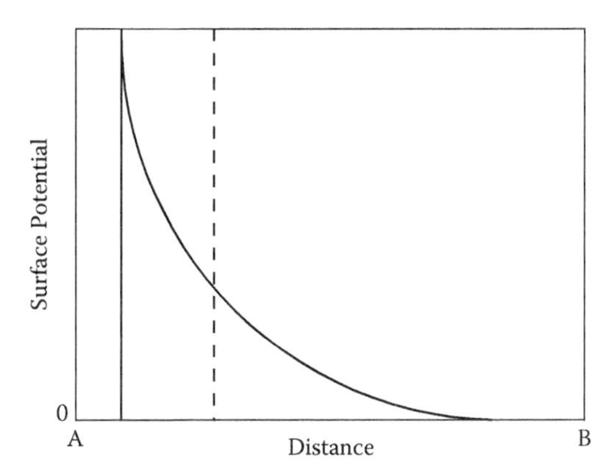

FIGURE 4.8 Double-layer model. (See text for details.)

The surface phase can be described by the Helmholtz double layer theory (Figure 4.8). If a negatively charged lipid molecule, R-Na+, is adsorbed at the interface AA′, the interface will be negatively charged (air–water or oil–water). According to the Helmholtz model for the double layer, Na+ on the interfacial phase will be arranged in a plane BB′ toward the aqueous phase. The distance between the two planes, AA′ and BB′, is given by Ù. The charge densities are equal in magnitude, but with opposite signs, Γ (charge per unit surface area), in the two planes. The (negative) charge density of the plane AA′ is related to the surface potential (negative), φ_o, at the Helmholtz charged plane:

$$\varphi_o = (4\pi\sigma\delta)/D \tag{4.17}$$

where D is the dielectric constant of the medium (aqueous). According to Helmholtz double layer model, the surface potential decreases sharply from its maximum value, φ_o, to zero as δ becomes zero (Birdi, 1989, 1999). The variation of surface potential is depicted in Figure 4.8.

It was found that the Helmholtz model was not able to give a satisfactory analyses of measured data. Later, another theory of the diffuse double layer was proposed by Gouy and Chapman. The interfacial region for a system with a charged lipid, R-Na+, with NaCl, is shown in Figure 4.9.

As in the case of the Helmholtz model, the plane AA′ will be negative due to the adsorbed R-species. Due to this potential, the Na^+ and Cl^- ions will be distributed nonuniformly due to the electrostatic forces. The concentrations of the ions near the surface can be given by the Boltzmann distribution, at some distance x from the plane AA′ as:

$$C^s_{Na+} = C_{Na+}\,(\exp - (\varepsilon\,\varphi/k\,T)) \tag{4.18}$$

$$C^s_{Cl-} = C_{Cl-}\,(\exp + (\varepsilon\,\varphi/k\,T)) \tag{4.19}$$

where C_{Na+} and C_{Cl-} are the number of sodium and chloride ions per milliliter, respectively, in the bulk phase. The magnitude of φ varies with x as shown in Figure 4.9,

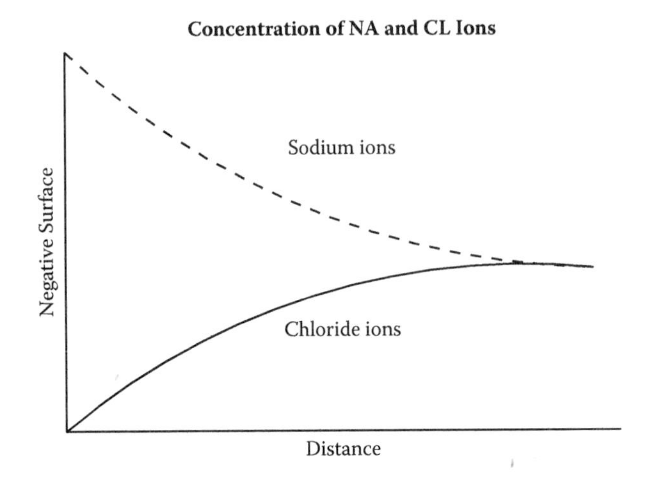

FIGURE 4.9 The variation of Na+ and Cl– ions near the charged surface (negative surface due to adsorbed R–).

from its maximum value, φ_o, at plane AA′. From the preceding equations we find that the quantities C^s_{Na+} and C^s_{Cl-} will decrease and increase, respectively, as the distance x increases from the interface, until their values become equal to C_{Na+} and C_{Cl-}, where φ is zero. The variations of C^s_{Na+} and C^s_{Cl-} with x are given in Figure 4.10. The extended region of x between AA′ and BB′ in Figure 4.8 may be termed the diffuse or the *Gouy-Chapman double layer.*

The volume density of charge (per ml) at a position within AA′BB′ may be defined as equal to:

$$\varepsilon\ (cs+ - cs-) \tag{4.20}$$

can be expressed by the Poisson relation:

$$d^2\ \varphi/d^2\ x = -(4\ \pi)/D \tag{4.21}$$

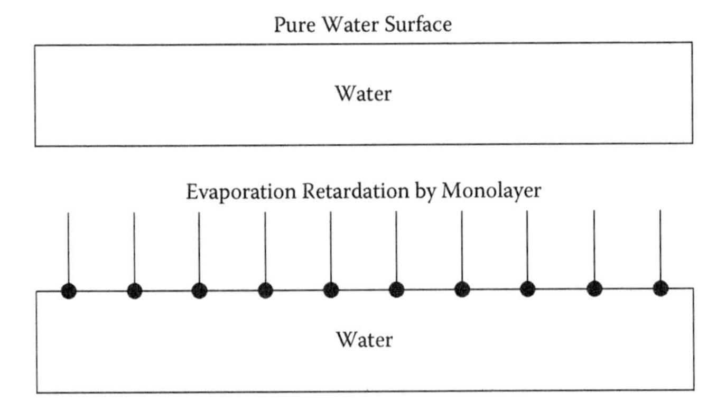

FIGURE 4.10 The effect of the lipid monolayer on the rate of evaporation of water from lakes.

In this derivation it is assumed that the interface is flat, such that it is sufficient to consider only changes in φ in the x direction normal to the surface plane.

The following relation can be derived from the above:

$$(d\ \varphi/d\ x) = d(d\varphi/dx)/dx = -(4\ \pi\ N\ \varepsilon/1000\ D)\ c \tag{4.22}$$

$$(e^{-\varepsilon/kT}\ e^{+\varepsilon\psi/kT}) \tag{4.23}$$

where c is the bulk concentration of the electrolyte.

In a circle of unit surface area on the charged plane A–A′, the negative charges acquired by the adsorbed organic ions (amphiphiles) within this area represents the surface charge density, σ:

$$\sigma = -\int \rho\ d\ x \tag{4.24}$$

$$= (D/4\ \pi)(d\ \varphi/d\ x) \tag{4.25}$$

when integration is zero and infinity.

At x = 0, the magnitude of φ reaches φ_o:

$$\sigma = (2DRTc/1000\ \pi)^{1/2}\ [\sinh\ (\varepsilon\ \varphi^o/2kT)] \tag{4.26}$$

The magnitude of the thickness of the double layer, $1/k$, that is, Debye-Huckel length, has been expressed as (Chattoraj and Birdi, 1984) follows:

$$1/k = (1000\ D\ R\ T/8\ \pi N\varepsilon c)^{0.5} \tag{4.27}$$

At 25°C, for uni-univalent electrolytes one gets:

$$k = 3.282\ 10^7\ c^{-1}\ (cm) \tag{4.28}$$

For small values of φ, one gets the following relationships:

$$\sigma = (DRT\ k/2\pi N\varepsilon)\ \sinh(\varepsilon\varphi_o/2kT) \tag{4.29}$$

This relates the potential charge of a plane plate condenser to the thickness $1/k$. The expression based upon the Gouy model is derived as:

$$\sigma = 0.3514\ 10^5\ \sinh\ (0.0194\ \varphi_o) \tag{4.30}$$

$$= \Gamma z N\varepsilon \tag{4.31}$$

where the magnitude of Γ can be experimentally determined, and the magnitude of φ_o can be estimated. The free energy change due to the electrostatic work involved in charging the double layer is (Chattoraj and Birdi, 1984; Birdi, 1989, 1999; Adamson and Gast, 1997):

$$Fe = \int \psi o\ \sigma\ d\ \varphi \tag{4.32}$$

By combining these equations one can write the expression for Π_{el} (Chattoraj and Birdi, 1984):

$$\Pi_{el} = 6.1\ c^{1/2}\ (\cosh \sinh^{-1}\ (134/A_{el}\ C^{1/2}) - 1) \quad (4.33)$$

The quantity k T is approximately 4 10^{-14} erg at ordinary room temperature (25°C), and k T/ε = 25 mV. The magnitude of Π_{el} can be estimated from monolayer studies at varying pH. At the isoelectric pH, the magnitude of Π_{el} will be zero (Birdi, 1989). These Π versus A isotherms data at varying pH subphase has been used to estimate Π_{el} in different monolayers. At the isoelectric pH, the value of Π_{el} was zero, as expected.

The role of ions in biology is one of the most important fields of research. It has been found that different ions (both positively and negatively charged) contribute to important biological functions. The most important biological cell membrane function is the transport of ions (such as Na, K, Li, Mg) through the hydrophobic lipid part of the bilayer lipid membrane (BLM) (Birdi, 1989, 1999). This property has relations to many diseases, such as infection and thus the activity of antibiotics. These complicated biological processes have been studied by using monolayer model systems.

For instance, valinomycin (a peptide, C54H90N6O18) monolayers have been extensively investigated. Valinomycin is a cyclic desipeptide antibiotic that breaks down cells by greatly increasing the permeability of membranes specifically to potassium ions. Valinomycin has a 1000 higher binding selectivity for K+ than Na+ ion. The monolayers have exhibited K-ion specificity, exactly as that found in the biological cells. As is well known, cell membranes inhibit the free transport of ions (the alkyl chains of the lipids hinder such transport). However, molecules such as valinomycin assist in specific ion (K-ion) transport through binding (Birdi, 1989). The molecule valinomycin has a specific binding site for K^+ ion, which gives rise to its biological application. In fact, an electrode coated with a layer of valinomycin behaves as a potassium ion electrode (available commercially).

4.3 EFFECT OF LIPID MONOLAYERS ON EVAPORATION RATES OF LIQUIDS

In arid and semiarid areas the amount of water lost from reservoirs and lakes by evaporation frequently exceeds the amount beneficially used. As is well known, loss of water by evaporation from lakes and other reservoirs is a very important phenomenon in those parts of world where water is not readily available. Further, from ecological considerations (rain and temperature) the evaporation phenomena has much importance for global temperature and other phenomena. Cloud formation depends on the evaporation of water from lakes and oceans. The clouds are thus isolators for sunlight reaching on earth. This cycle is thus related to the global heating process. This has been extensively discussed with regard to CO_2 and global heating of the earth. The reduction of even just a part of the evaporation losses would therefore be of incalculable value for climate control.

The amount of loss of water from the surface will be expected from different processes:

Fall in water level + Rainfall = Evaporation + Seepage + Abstractions

Lipid monolayers are found to have extensive affect on the evaporation rates of water. In one example:

Surface area of water = 60 cm^2
Rate of evaporation for pure water surface = 0.66 mg/sec
Rate of evaporation with a stearic acid monolayer = 0.34 mg/sec

---

At the air–water interface the water molecules are constantly evaporating and condensing in a closed container. In an open container the water molecules at the surface will desorb and diffuse into the gas phase. It thus becomes of importance to determine the effect of a monomolecular film of amphiphile at the interface. Measurement of the evaporation of water through monolayer films was of considerable interest in the study of methods for controlling evaporation from great lakes. Many important atmospheric reactions involve interfacial interactions of gas molecules (oxygen and different pollutants) with aqueous droplets of clouds and fogs, as well as ocean surfaces. The presence of monolayer films would thus have an appreciable effect on such mass transfer reactions.

In the original procedure, the box containing the desiccant is placed over the water surface, and the amount of water sorbed is determined by simply removing the box and weighing it (Birdi, 1989; Adamson and Gast, 1997). The results are generally expressed in terms of specific evaporation resistance, r. The methods for calculating resistance from the water uptake values, together with the assumptions involved, are described in detail in the aforementioned references. The rates of evaporation are measured both without (R_w) and with (R_f) the film. The resistance is given as:

$$r = A\,(v_w - v_d)\,(R_f - R_w) \tag{4.34}$$

where A is the area of the dish, and v_w and v_d are the water vapor concentrations for water and desiccant, respectively. The condensed monolayers gave much higher r values than the expanded films, as expected.

A more advanced arrangement was described that made use of recording film balances, in which the temperature difference between two cooled sheet metal probes was measured. If both probes are over clean water and the rate of moisture condensation is the same, then there is no difference in the temperature. However, if a monolayer is present, the retardation of evaporation gives rise to a temperature difference.

Other techniques where the diffusion of gas through monolayers at the liquid interface have also been investigated. In these methods a differential manometer system was used to measure the adsorption of gases, such as CO_2 and O_2, into aqueous solutions with and without the presence of monolayers. The use of Geiger-Mueller counter with a suitable sorbent and a radioactive tracer gas was used to measure the reduction of evolution of H_2S and CO_2 from surface solution when a monolayer was present. Many decades ago it was found that the evaporation rates of water was reduced by the presence of monomolecular lipid film (Figure 4.10). This observation gave rise to important application in reducing the loss of water from great lakes during summer in such countries as Australia and Africa.

The simplest method to use to investigate is to measure the loss of weight of a water container, with or without the presence of a monomolecular lipid film. La Mer investigated the effect of long chain films (C14 H29 OH, C16 OH, C18 OH, C20 OH, C220H) and found that the resistance to evaporation increased with the chain length (Adamson and Gast, 1997). For instance, the resistance increased by a factor of 40 for C22OH as compared to C14OH. This indicates that evaporation takes place mainly through the alkyl chains films. Since the attraction between alkyl chains increases with number of carbon atoms, then the resistance to evaporation increases.

Mixed lipid monolayers provide the packing and orientation of such molecules at the water interface. These interfacial characteristics affect many other systems. For instance, one uses mixed surfactants in froth flotation. The monolayer surface pressure of a pure surfactant is measured after the injection of the second surfactant. From the change in surface pressure one can measure the interaction mechanism. The monolayer method has also been used as a model biological membrane system. In the latter bilayer lipid membranes (BLMs), lipids are mixed with other lipid-like molecules (such as cholesterol) (Birdi, 1999). Hence, mixed monolayers of lipids plus cholesterol have been found to provide much useful information on BLM. The most important BLM and temperature melting phenomena is the human body temperature regulation. Normal body temperature is 37°C (98°F) at which all BLM function efficiently.

In biological systems, the lung fluid exhibits surface pressure characteristics that are related to its lipid composition. The ratio between two different lipids has been shown to be critical for lung function in newborn babies. Recent studies of mixed monolayers have been made by using AFM (Birdi, 2002).

4.4 MONOLAYERS OF MACROMOLECULES AT THE WATER SURFACE

It is already obvious that monolayers on water are only stable if the hydrophobic part of the molecule is of the right magnitude as compared to the polar part. Many macromolecules (such as proteins) form stable monolayers at water surface if the hydrophilic–lipophilic balance (HLB) is of the right quantity. Especially, almost all proteins (hemoglobin, ovalbumin, bovine serum albumin, lactoglobulin, etc.) are reported to form stable monolayers at water surface (Birdi, 1999).

4.5 LANGMUIR-BLODGETT (L-B) FILMS (TRANSFER OF LIPID MONOLAYERS ON SOLIDS)

It is obvious that the lipid monolayers as spread on water surface are not easily visualized at molecular scale. Some decades ago (Langmuir, 1920) it was reported that when a clean glass plate was dipped into water covered by a monolayer of oleic acid that an area of the monolayer equal to the area of the plate dipped was deposited on withdrawing the plate. Later, it was found that any number of layers could be deposited successively by repeated drippings, which was called the Langmuir-Blodgett (L-B) method (Birdi, 2002, 2009).

In another context, the electrical properties of thin films obtained by different procedures, for example, thermal evaporation in vacuum, have been investigated in much detail. However, the films deposited by the L-B technique have only recently been used in the electrical applications. The thickness in L-B films can be varied from only one monomolecular layer (25 Å = 25 10^{-10} m), while this is not possible by the evaporation procedures. Monomolecular layers (L-B films) of lipids are of interest in a variety of applications including the preparation of very thin controlled films for interfaces in solid-state electronic devices (Gaines, 1966; Birdi, 1989, 1999). The process of transferring the spread monolayer film to a solid surface by raising the solid surface through the interface has been studied for many decades. This process of transfer is depicted in Figure 4.11. It is seen that if the monolayer is a closely packed state, then the monolayer is transferred to the solid surface, most likely without any change in the packing density. Detailed investigations have, however, shown

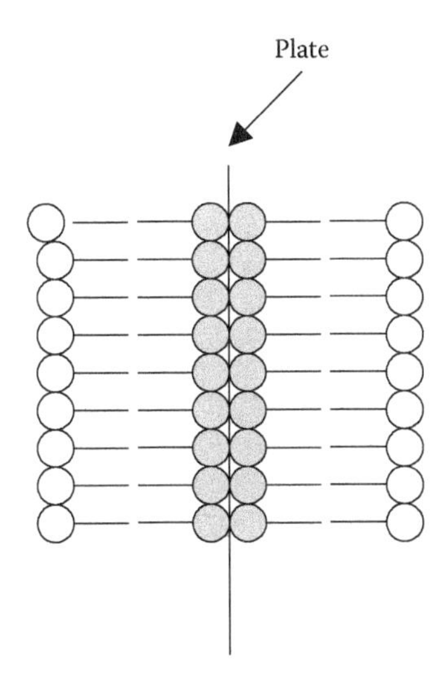

FIGURE 4.11 Transfer of monolayers of lipids to L-B films.

that the process of transfer is not as simple as shown in Figure 4.11. The monolayer on a solid surface may or may not be stable, and the defects may be present. However, by using the modern AFM techniques, one can determine the molecular orientation and packing of such L-B films (Gaines, 1968; Birdi, 1989, 1999, 2002).

L-B film technology is rapidly developing in such areas as:

- Microlithography
- Solid-state polymerization
- Light guiding
- Electron tunneling
- Photovoltaic effects

In the case of such films as Mg-stearate, if a clean glass slide is dipped through the film, a monolayer is adsorbed on the downstroke. Another layer is adsorbed on the upstroke. Under careful conditions, one may make L-B films with multilayers (varying from a few layers to over thousands). One can monitor the adsorption by measuring the decrease in surface pressure on each stroke. If no adsorption takes place, then no change in surface tension is observed. There are some lipids, such as cholesterol, that do not form L-B films. There are also other methods that can provide detailed information on L-B films, such as:

- Light reflection
- IR spectroscopy
- STM (scanning tunneling microscopy) and AFM (atomic force microscopy)
- Change in contact angle

The L-B deposition is traditionally carried out in the monolayer in a solid state. The surface pressure is then high enough to ensure sufficient cohesion in the monolayer, for example, the attraction between the molecules in the monolayer is high enough so that the monolayer does not fall apart during transfer to the solid substrate. This also ensures the buildup of homogeneous multilayers. The surface pressure value that gives the best results depends on the nature of the monolayer and is usually established empirically. However, amphiphiles can seldom be successfully deposited at surface pressures lower than 10 mN/m, and at surface pressures above 40 mN/m collapse, and film rigidity often poses problems. When the solid substrate is hydrophilic (glass, SiO_2, etc.) the first layer is deposited by raising the solid substrate from the subphase through the monolayer, whereas if the solid substrate is hydrophobic (HOPG, silanized SiO_2, etc.) the first layer is deposited by lowering the substrate into the subphase through the monolayer. In some L-B systems, the magnitude of surface pressure drops about 1 to 2 mN/m each time a plate (with a surface area of 5 cm^2) is moved down or up through the monolayer. It is thus a very sensitive method to study the L-B deposition phenomena directly. There are several parameters that affect what type of L-B film is produced. The quantity and the quality of the deposited monolayer on a solid support are measured by a so-called transfer ratio, t_r. This is defined as the ratio between the decrease in monolayer area during a deposition stroke, Al, and the area of the substrate, As. For ideal transfer the magnitude of t_r is equal to 1. Depending on the behavior of the molecule the solid substrate can be dipped through the film until the desired thickness of the film is achieved. Different kinds of L-B multilayers can be produced or obtained by successive deposition of monolayers on the same substrate (see Figure 4.11). The most common one is the Y-type multilayer, which is produced when the monolayer deposits to the solid substrate in both up and down directions. When the monolayer deposits only in the up or down direction the multilayer structure is called either Z-type or X-type. Intermediate structures are sometimes observed for some L-B multilayers and they are often referred to be XY-type multilayers.

The production of so-called alternating layers, which consist of two different kinds of amphiphiles, is also possible by using suitable instruments. In such an instrument there is a trough with two separate compartments both possessing a floating monolayer of a different amphiphile. These monolayers can then be alternatingly deposited on one solid substrate.

The monolayer can also be held at a constant surface pressure, which is enabled by a computer-controlled feedback system between the electrobalance and the motor responsible for the movements of the compressing barrier. This is useful when producing L-B films, that is, when the monolayer is deposited on a solid substrate. One of the simplest procedures generally used is where a clean and smooth solid surface (of suitable surface area) is dipped through the interface with the monolayer. Alternatively, one can also place the solid sample in the water before a monolayer is spread, and then draw up through the interface to obtain the film transfer.

It is obvious that such processes involving monomolecular film transfers will easily be disturbed by defects, arising from various sources. As will be shown later, these defects are in most cases, easily detected. The structural analysis of the molecular ordering within a single L-B monolayer is important both to understand how the environment in the immediate vicinity of the surface (i.e., solid) affects the structure of the molecular monolayer, and to ascertain how the structure of one layer forms a template for subsequent layers in a multilayer formation. Studies of the order within surfactant monolayers have been reported for many decades. Multilayer assemblies have been studied by electron as well as infrared absorption.

4.5.1 Electrical Behavior of L-B Films

Insulating thin films in the thickness range 100 to 20,000 Å (100 to 20,000 10^{-10} m) have been a subject of varied interest among the scientific community because of the potential applied significance for developing devices, such as optical, magnetic, and electronic. Some of the unusual electrical properties possessed by thin L-B films, which are unlike those of bulk materials, led to the possibility of their technological applications and, consequently, interest in thin film studies grew rapidly. Earlier studies did not prove to be very inspiring because the L-B films obtained always suffered with the presence of pinholes, stacking faults, and other impurities, and hence the results were not reproducible. It is only in the past few decades that many sophisticated methods have become available for the production and examination in thin films, and reproducibility of the results could be controlled to a greater extent. Nevertheless, the unknown nature of inherent defects and a wide variety of thin film systems still complicates the interpretation of many experimental data and thus hinders their use in devices. It has been found that the breakdown conduction in thin films, the major subject of investigation have been based on the films prepared by thermal evaporation under vacuum or similar techniques. It was realized that the L-B films have remained less known among the investigators of this field. The various interesting physical properties of L-B films have been investigated in the current literature. The L-B films being very sensitive assemblies, it is necessary that these structures are perfect. There are two crucial factors for making satisfactory electrical measurements on the L-B films: uniform packing and thickness. These two requirements are generally fulfilled to a greater extent when one uses L-B films. Metrical thicknesses of several fatty acid monolayers have been measured using the best known optical methods. By employing a protective colloid ion layer the thickness was measured, which had a high quality of fringes and accordingly high precision of measurements.

With regard to the uniformity in the packing of these L-B films, it is known that the selection and purity of substrates needs to be carried out very carefully. However, it has been reported that increasing the number of monolayers increases the degree of uniformity of the film, and this observation has been confirmed by other investigators. In fact to achieve the required uniformity in the packing of L-B films, the deposition of the first monolayer has been critical, since any voids or imperfections in it may generally lead to a major disruption of the subsequent monolayers. Therefore, it has been suggested that the solid slide emerging from the liquid surface must be completely dried before being reimmersed. If the wet slide is reimmersed, generally, no multilayers are adsorbed (for lecithin, fatty acids, and protein monolayers). L-B films of Ba- or Cu-stearate have been built up to as many as 3000 layers simply by repeated dipping and withdrawal processes. On the other hand, fatty acid films or another source give rise to cracking and fogging tendencies. It was suggested that the addition of Cu ion [10^{-6} M] did not give rise these difficulties (Birdi, 1999).

4.5.2 Optical and Related Methods of Optical Absorption

The optical reflection from a monomolecular film must be measured from the interface with a very small amount of material present. Therefore, in these methods repeated interaction of the light beam with one or more identical films is generally used. The simplest way to observe a light beam that has passed through several identical monolayers is to transfer portions of the layer to suitable transfer end plates, which are then stacked and examined in a conventional spectrophotometer. This method was used where monolayers of chlorophyll were deposited on glass slides by the L-B method. Ferrodoxin and chlorophyll monolayers were investigated by measuring the spectra (550–750 nm) of these films at the interface.

4.5.3 L-B Deposited Film Structures

No direct method exists by which one can study the monolayer film molecular structures on water (i.e., *in situ*). Therefore, the L-B method has been used to study molecular structures in the past decades. The most useful method for investigating the detailed L-B deposited film structure is the well-known electron diffraction technique (or the scanning probe microscopes) (Birdi, 2003). The molecular arrangement of deposited mono- and multilayer films of fatty acids and their salts using this technique have been reported. These analyses showed that the molecules were almost perpendicular to the solid surface in the first monolayer. It was also reported that Ba-stearate molecules have a more precise normal alignment than in the case of stearic acid monolayers. In some investigations, thermal stability of these films has been remarkably stable, up to 90°C.

Based on these structural analyses obtained by the electron diffraction technique, these deposited films are known to be monocrystalline in nature, and thus can be regarded as a special case of a layer–bilayer mechanical growth forming almost "two-dimensional" crystals. There is, however, evidence that Ba-behenate multilayers do in fact show an absence of crystallization, which has been demonstrated by electron micrographic studies.

In many of the early reports, it was shown that deposited films obtained by the usual process of monolayer transfer invariably contained holes, cracks, or similar imperfections. These observations are not surprising because at higher surface pressure, the more compact film will be removed. Nevertheless, it would be an oversimplification to regard the film transferred at high surface pressure as perfectly uniform, coherent, and defect free. In a recent study the L-B films of iron stearate was deposited from aqueous solution into a substrate only on down-journey. On up-journey the substrate was withdrawn through a clean area. The condition of aqueous surface was the same as for the preparation of y-type layer. Fe55-tracer techniques were used for the examination of the direction of molecules. The unidirectional surface conductance of monolayers of stearic acid deposited on a glass support was investigated. The contact angles and adhesional energy changes during the transfer of monolayers from the air–water interface to solid (hydrophobic glass) supports has been analyzed (Gaines, 1966; Birdi, 1989).

4.5.4 Molecular Orientation in Mixed Dye Monolayers on Polymer Surfaces

During the last few years, there has been an increasing interest in the use of surface-active dyes to study properties of biological and artificial membranes and to construct monomolecular systems by self-organization. When these dyes are incorporated in lamellar systems, it has been found that the paraffin chains stand perpendicularly on the plane of the layer while the chromophores lie flat near the hydrophilic interface. In order to develop new molecules as functional components of monolayer assemblies a series of nine surface-active azo and stilbene compounds were synthesized. Their monolayer properties at the air–water interface were investigated by surface pressure and spectroscopic techniques. The adsorption on stretched polyvinyl alcohol (PVA) films of mixed monolayers of n-octadecyltrichlorosilane (OTS-C18H37SiCl3) plus long chain substituted cyanine dyes were investigated. These systems were selected due to:

The solid support possesses a simple organization with uniaxial symmetry, which allows a straightforward correlation with the molecular orientation induced in the monolayer.

The solid substrate (PVA) is transparent and the adsorbed molecules contain elongated dye chromophores, so that the molecular orientation within the monolayer can be readily determined by means of polarized absorption spectroscopy (linear dichroism, ld).

It is easy to produce the support in large quantities, and adsorption is easy for oleophobic OTS monolayers. Furthermore, PVA is not soluble and does not swell in the organic solvents used in monolayer studies. The orientational effects were estimated from linear dichroism.

Interest in the dielectric studies of deposited L-B films of fatty acids and their metal salts was one of the parameters of main investigations in the early stages of research on L-B films, for example, capacitance, resistance, and dielectric constant. In early investigations, measurements on impedance of films and related phenomena were carried on Cu- and Ba-stearate and Ca-stearate using both x- and y-type films. Initially, Hg droplets were used for small area probe measurements and an alternating current (AC) bridge was used for impedance measurements. The resistance of the films was very low (<1 ohm) with high signal voltages, whereas it was of the order of megaohms with signals of 1 or 2 volts. In both types of films, the capacity decreased with thickness, as can be expected from the following relation:

Capacitance of the deposited L-B films:

$$C_C = A\ \varepsilon/4\ \pi\ N\ d \tag{4.35}$$

in which C_C is the capacity, ε denotes the dielectric constant, N is the number of layers, d the layer thickness, and A the area of contact between drop and film. On the other hand, the values of the resistance per layer showed a definite increase with the thickness of the film. The specific resistance of the films, thus determined from their values of the resistance per layer was about 10^{13} ohms. This was based on the results of capacity measurements on some 75 samples. The capacitance measurements thus performed on stearate films (1–10 layers) led to ε values between 2.1 and 4.2, with a bulk value of 2.5.

In many of the measurements reported in the literature, the organic film was sandwiched between evaporated aluminum electrodes. The fact that an oxide layer grows on the base of an aluminum electrode was present and its effect on the capacitance values of the device was neglected considering that the resistivity of oxide film is small compared with resistivity of the organic L-B layers. The presence of such a thin oxide layer between metal electrodes and fatty acids can be analyzed. The capacitance has been reported to be a linear function of $1/C_C$ with respect to the number of transferred monolayers (Figure 4.12). L-B films of Ba-salts of fatty acids deposited at Π = 50 mN/m (Birdi, 2003), gave following relation between [$1/C_C$] and N:

$$\text{Ba-stearate: } 1/C_C = 15.9\ N + 1.13,\ (10^6\ F^{-1}) \tag{4.36}$$

$$\text{Ba-behenate: } 1/C_C = 17.2\ N + 8,\ (10^6\ F^{-1}) \tag{4.37}$$

The dielectric anisotropy of long-chain fatty acid monolayers was analyzed. These fatty acids were considered as being oriented in a cylinder cavity with length (L) >>diameter (D). Considering each bond in these molecules as a polarization ellipsoid with axial symmetry about the -C-C- bonds, the mean polarizability of the bonds was calculated.

The L-B films of fatty acid salt are known to have breakdown strength, about 10^6 V/cm. The earliest study of breakdown of the films under high direct current (DC) voltages were measured by using a galvanometer to read the current. By using an Hg drop as the upper electrode, no current was detected until the magnitude of the applied voltage across the specimen reached critical value in the case of a 30-layer L-B film, whereas in thinner films appreciable currents were present almost from the start. The breakdown per layer was found to rise sharply at a thickness of about 20 layers when its order of magnitude was about 10^6 V/cm. The specific resistance was approximately 10^{13} ohms below the breakdown voltages. In later studies, water drop with electrolytes was used instead of Hg, and no difference was found between the two methods. The destructive breakdown in L-B films was found to be composed of two events: (1) the destructive breakdown voltage and (2) the maximum breakdown voltage. These events have been determined in various fatty acid soaps in the thickness range corresponding to 16 to 80 layers. The destruction breakdown voltage has

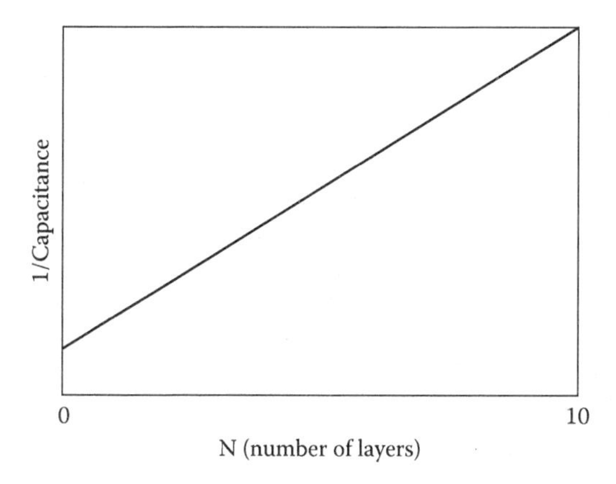

FIGURE 4.12 Variation of 1/C (C = capacitance) of L-B films versus the number of layers (N).

been reported to be independent of the material and thickness of the film. But the maximum breakdown voltage has been reported to vary slightly with an increase film thickness. The breakdown mechanism in L-B film has been described in terms of a statistical model (Agarwal, 1974; Birdi, 1999).

The study of c-voltage characteristics in metal-oxide semiconductor (MOS) structures has provided very useful information about the mechanism of certain types of devices. The L-B films of Ca-behenate have been investigated with orthophenanthroline with three stearate chains. In these studies the p-type silicon surfaces (metal-insulator-semiconductor, MIS), after etching in CP4, cleaning, and drying, were used to deposit L-B films, and were covered with aluminum in the form of two circles connected by a channel. From the linear relation between 1/C versus N, the thickness was estimated to be 15 Å. A small variation in BIAS voltage gave rise to a large modulation of c, which has been ascribed to the small thickness of the insulator and from the high resistivity (ca = 600 Ω cm) of the p-type silicon. So far only polymerizable sites have been described, and monomolecular layers involving radiation-cleavageble sites appear unreported. One approach to such a site is the inclusion of a quaternary center or a heavy element in the chain to enhance electron beam and x-ray cross-section.

The transfer rate of dipalmitoyl lecithin (DPPC) monolayer from aqueous phase to glass surface has been investigated. The change in contact angle of DPPC monolayer–glass interface, where DPPC was kept at fixed area/molecule, 134 Å, was reported to be sigmoidally increased. These rates were analyzed as first-order kinetics, and the transfer rate was 2.59 10^{-3} 1/sec.

The interactions between insoluble monolayers of ionic surfactants and ions dissolved in the subsolution have been an object of continuing scientific interest (Gaines, 1966; Birdi, 1989; Adamson and Gast, 1997). This interest rests upon the desire to model important natural and technological processes such as enzymatic activity, permeability of cellular membranes, ion exchange processes in soils, adsorption in ion exchange resins, ion flotation, and chromatography. Stearic acid monomolecular films on 10^{-3} M $CaCl_2$ subsolutions were deposited in paraffin-coated microscope glass slides by the L-B technique (pH range 2 to 9).

The combination of the L-B technique and the neutron activation method of analysis was used to determine the stoichiometry of the interaction between fatty acid (arachidic acid) and metal ions dissolved in the subphase. The experimental data were used to estimate the stability constants of arachidic acid and bivalent metal ions (Cd and Ba). The data were explained as an interaction between metal ions and monolayer as an adsorption process:

$$2\ RCOO^- + Me^{++} <=> (RCOO)^2\ Me \qquad (4.38)$$

The data for various binding constants have been reported in the literature. These data were in good agreement with other literature studies.

Physical properties of L-B films: The surface property of a solid changes even after one layer of a lipid is formed as in an L-B film. For example, the contact angle decreases after L-B deposition (in addition to other properties). Similarly, many other physical methods have shown that L-B films change the surface characteristics (Gaines, 1966; Adamson and Gast, 1997; Birdi, 2002, 2003).

Fourier transform infrared (FTIR)-attenuated total reflection (ATR) spectra have been investigated of L-B films of stearic acid deposited on a germanium plate with one, two, three, five, and nine monolayers. The C = O stretching band at 1702 cm^{-1} was missing for the monolayer. The intensity increased linearly for the multilayer samples. CH scissoring band at 1468 cm^{-1} appeared as a singlet in the case of 1-monolayer. A doublet at 1473 and 1465 cm^{-1} was observed for films containing more than three monolayers. Band progression due to -CH2- wagging vibration of the trans-zigzag hydrocarbon chains is known to appear between 1400 cm^{-1} and 1180 cm^{-1}. The intensities increased in this region with the number of layers (Birdi, 1999).

4.5.5 Application of L-B Films in Industry

Thus, the L-B film is seen to be a sensitive structure where a layer of lipid can be arranged on any solid. The applications of such structures in various industrial applications has been a recently exploited area. Since even a single monolayer of lipid on a solid surface (such as glass, metal, etc.) results in a large change in the contact angle of water, this indicates the potential of such a procedure. Thin films of phthalocyanine compounds, in general, and those prepared by the L-B method, in particular, display novel electrical properties (Birdi, 1999). The L-B technique for depositing mono- and multilayer coatings with well-controlled thickness and morphology offers excellent compatibility with microelectronic technology. Such films have recently been reviewed for their potential applications. The combination of L-B supramolecular films with small dimensionally comparable microelectronic substrates affords new opportunities for generation of fundamental chemical property information and evaluation of new organic thin film semiconductors as microelectronic components and devices.

4.6 BILAYER LIPID MEMBRANES (BLMs)

Scientists have been engaged in determining the molecular structures of biological membranes. However, it was realized that the biological cells were contained by some kind of a thin *lipid membrane*. Some decades ago one did not have any experimental technique that allowed for seeing the molecular structure of cells. In order to analyze this in more detail, experiments were made (a few decades ago) as follows. Lipids were extracted from the biological cells. These lipids were compressed on the Langmuir balance and the value of area per molecule was estimated (about 45 $Å^2$/molecule). Knowing the diameter of the cells and from the amounts of lipids (and the area/molecule data), one reached the conclusion that the cell membranes were composed of a *bilayer of lipids* (bilayer lipid membrane, BLM). This was one of the most important results in the history of biological cell membranes. Later, these results were confirmed from x-ray diffraction data and other scanning probe microscopes (SPMs) (Birdi, 1999, 2002; Woodson and Liu, 2007).

4.7 DIVERSE APPLICATIONS

4.7.1 Vesicles and Liposomes

It was mentioned that ordinary surfactants (soaps, etc.) when dissolved in water form self-assembly micellar structures. The phospholipids are molecules like surfactants; they also have a hydrophilic head and generally have two hydrophobic alkyl chains. These molecules are the main components of the membranes of cells. The lung fluid also consists mainly of lipids of this kind. In fact, usually the membranes of cells are made up of two layers of phospholipids, with the tails turned inward, in the attempt to avoid water. The external membrane of a cell contains all the organelles and the cytoplasm.

Phospholipids when dispersed in water may exhibit self-assembly properties (either as micellar self-assembly aggregates or some larger structures). This may lead to aggregates that are called liposomes or vesicles (Reimhult et al., 2003; Birdi, 2008). Liposomes are structures that are empty cells and are currently being used by some industries. They are microscopic vesicles or containers, formed by the membrane alone. They are widely used in the pharmaceutical and cosmetic fields because it is possible to insert chemicals inside them. Liposomes can also be used to solubilize (inside the hydrophobic part) hydrophobic chemicals (water-insoluble organic compounds) such as oily substances so that they can be dispersed in an aqueous medium by virtue of the hydrophilic properties of the liposomes (in the alkyl region).

A certain type of lipid (or lipid-like) molecule, when dispersed in water, tends to make self-assembly structures (Figure 4.13). Detergents have been shown to

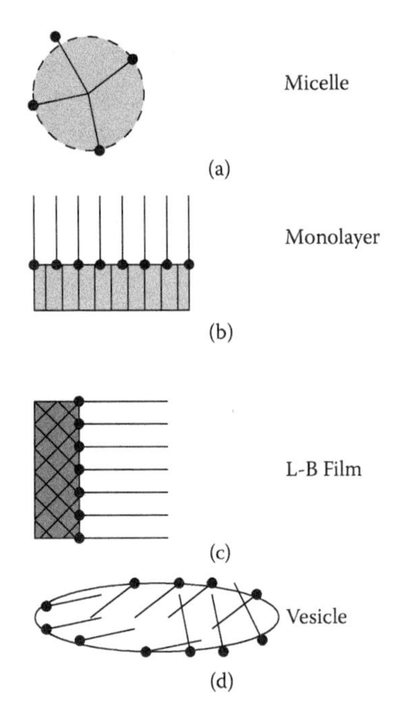

FIGURE 4.13 Different lipid self-assembly monolayer (SAM) structures: (a) micelle; (b) monolayer; (c) L-B film; (d) vesicle.

aggregate to spherical or large cylindrical-shaped micelles. It is known that if egg phosphatidylethanolamine (egg lecithin) is dispersed in water at 25°C, it forms a self-assembly structure, which is called liposome or vesicle. A liposome is a spherical vesicle with a membrane composed of a phospholipid and cholesterol (less than 50%) bilayer. Liposomes can be composed of naturally-derived phospholipids with mixed lipid chains (such as egg phosphatidylethanolamine), or of pure surfactant components like DOPE (dioleolylphosphatidylethanolamine). Liposomes usually contain a core of aqueous solutions. Multilayer liposomes are called vesicles. However, a range of mixtures of these structures in mixed lipid systems is found.

In all these structures, the main characteristic is due to self-assembly properties of the lipid molecules. Hence, it is safe to conclude that many important natural systems are based upon this molecular characteristic of lipids. Vesicles are *unilamellar* phospholipid liposome. *Liposome* comprises two words from Greek: *lipid*, which means "fat," and *soma*, which means "body." The word *liposome* does not in itself denote any size characteristics. Furthermore the term *liposome* does not necessarily mean that it must contain lipophobic contents, such as water, although it usually does. The vesicles may be conceived as microscopic (or nanosized) containers carrying molecules (drugs) from one place to another. The structures are suitable for both transporting water-soluble or water-insoluble drugs. Since the lipids used are biocompatible molecules, this may also enhance their adsorption and penetration into cells. During the past decades, liposomes have been used for drug delivery due to their unique solubilization characteristics for water-insoluble organic substances. A liposome encapsulates a region on aqueous solution inside a hydrophobic membrane; dissolved hydrophilic solutes cannot readily pass through the lipids. Hydrophobic chemicals can be dissolved into the membrane, and in this way liposomes can carry both hydrophobic molecules and hydrophilic molecules. To deliver the molecules to sites of action, the lipid bilayer can fuse with other bilayers, such as the cell membrane, thus delivering the liposome contents. By making liposomes in a solution of DNA or drugs (which would normally be unable to diffuse through the membrane), they can be (indiscriminately) delivered past the lipid bilayer.

Liposomes can also be designed to deliver drugs in specific ways. Liposomes that contain low (or high) pH can be constructed such that dissolved aqueous drugs will be charged in solution (i.e., the pH is outside the drug's pI range) (isoelectric pH = pI). As the pH naturally neutralizes within the liposome (protons can pass through a membrane), the drug will also be neutralized, allowing it to freely pass through a membrane. These liposomes work to deliver drug by diffusion rather than by direct cell fusion. Another strategy for liposome drug delivery is to target endocytosis events. The size range of liposomes can be made such that these are suitable for viable targets for natural macrophage phagocytosis. In this way liposomes are consumed while in the magrophage's phagosome. In this process, the drug is target oriented.

Further, another important property of liposomes is their natural property to target cancer cells. The endothelial wall of all healthy human blood vessels are encapsulated by endothelial cells that are bound together by tight junctions. These tight junctions block large particles in the blood from leaking out the vessel. It is known that tumor vessels do not contain the same level of seal between cells and are diagnostically leaky. The size of liposomes can be used to play a specific application. For example, liposomes of certain sizes, typically (i.e., less than 400 nm), can rapidly enter tumor sites from the blood but are kept in the bloodstream by the endothelial wall in healthy tissue vasculature. Liposome-based anticancer drugs are now being used as drug delivery systems.

Liposomes can be created by shaking or sonicating phospholipids (dissolved in alcohol) in water. Low shear rates create *multilamellar liposomes*, which have many layers like an onion. Continued high-shear sonication tends to form smaller *unilamellar liposomes*. In this technique, the liposome contents are the same as the contents of the aqueous phase. Sonication is generally considered a "gross" method of preparation, and newer methods such as extrusion are employed to produce materials for human use. Further advances in liposome research have been able to allow liposomes to avoid detection by the body's immune system. These liposomes are known as *stealth liposomes* and have been made with PEG (polyethylene glycol) studding the outside of the membrane. The idea here has been as follows. The PEG coating is found to exhibit neutral properties in the body. In other words, due to electroneutrality, these kinds of liposomes do not attach to cells and can circulate in the system for much longer periods. Thus, the latter will show a longer circulatory life for the drug delivery mechanism. The specific targeting is also used in addition to the the PEG coating in stealth liposomes. This is achieved by attaching a specific biological species to the liposome. This enables binding to the targeted drug delivery site. In the literature one finds that different targeting ligands have been used (monoclonal antibodies, vitamins, or specific antigens). The target-oriented liposomes are thus useful, since drugs (some drugs are indeed toxic at high concentrations) are delivered to the tissues where needed. Further, it has been reported that liposomes can be covered with ligands to activate endocytosis in other cell types.

4.7.2 Applications in Drug Delivery

Drugs are used to cure a particular disease as related to a specific body function (such as the liver, lungs, eyes, heart, kidneys). Drugs are used in the form of tablets or injection or inhaling. A drug designed to cure liver or lung disease must reach its target with a suitable concentration. The main object is to treat the illness in any particular organ and the drug dosage is determined accordingly. However, if the drug breaks down in the process of transport through, for example, the stomach, then other innovations are needed. In the following, an example is given where drug delivery is designed through the nasal pathway.

There are at present many drugs that are applied through the nasal pathway (inhalable drug delivery, IDD). In addition to small molecules (such as hormones) even much larger molecules (such as insulin and other proteins) have been reported to be useful IDD systems. The inhale drug delivery needs to meet many critical demands:

- Deliver the drug effectively and reach the lungs
- The particles (in the form of aerosols) need to be designed to achieve consistent delivery
- Quantitative delivery of dosage

Surface active substances also need to be added, which are known to enhance the penetration through the skin barrier. The latter substances should not cause any irritation in the nose and other air pathways. Insulin is currently being commercially marketed as an IDD.

4.7.3 Diagnostics (Immobilized Enzymes on Solid Surfaces) Techniques and Surface Chemistry

During the past decade, extensive diagnostic instruments have become available to determine the state of different illness control. For instance, the concentration of glucose (in the case of diabetes control) in blood can be easily measured today by using a strip (1 mm × 1 mm) covered with a suitable enzyme (glucose oxidase), which in contact with blood sample reacts with glucose (within 30 sec) to produce degraded substances of glucose (production of hydrogen peroxide). This enzyme is very specific for the degradation of glucose. This reaction is calibrated to produce an electrical signal (glucose concentration varies from 3 to 30 mmol/L). Cholesterol can also be monitored by an enzyme that reacts specifically with cholesterol in the blood. The preparation of the diagnostic strip requires an even layer of the enzyme (or any other suitable chemical) on the test strip. These can be controlled by using the surface chemistry principles:

- Contact angle
- Surface tension of the applied solution
- Use of AFM to make image analyses

5 Solid Surfaces: Adsorption and Desorption of Different Substances

5.1 INTRODUCTION

In everyday life one finds that the surface of a material (glass, wood, plastics, metal, marble, gold or silver, etc.) plays an important role. One finds a large variety of applications where the *surface* of a solid has a specific role and function (for example, active charcoal, talc, cement, sand, catalysis). Solids are rigid structures and resist any stress effects. Thus it is seen that many such considerations in the case of solid surfaces will be somewhat different than those for liquids. The surface chemistry of solids is extensively described in the literature (Adamson and Gast, 1997; Zhuravlev, 2000; Birdi, 2002, 2009, 2010; Diebold, 2003; Neumann, 2010). The mirror polished surfaces of metals and plastics (such as in electrical appliances) are found to be of much importance in industry. Further, the *corrosion* of metals initiates at surfaces, thus requiring treatments that are based upon surface properties. In economic terms the process of corrosion is the most expensive. Surface treatment technology is constantly developing methods to combat corrosion, especially in cars, bridges, housing, and steel structures. As described in the case of liquid surfaces, analogous analyses of solid surfaces can be carried out. The molecules at the solid surfaces are not under the same force field as in the bulk phase (Figure 5.1).

The differences between perfect crystal surfaces and surfaces with defects are very obvious in many everyday observations. The solids were the first materials that were analyzed at molecular scale. This led to the understanding of the structures of solid substances and the crystal atomic structure. This is because although molecular structures of solids can be investigated by such methods as x-ray diffraction, the same analyses for liquids are not that straightforward. These analyses have also shown that surface defects exist at the molecular level.

As pointed out for liquids, it is important to consider that when the surface area of a solid powder is increased by grinding, then surface energy is needed. Of course, due to the energy differences between solid and liquid phases, these processes will be many orders of magnitude different from each other. Liquid state retains some structure that is similar to its solid state, but in liquid state the molecules exchange places. The average distance between molecules in the liquid state is roughly 10%

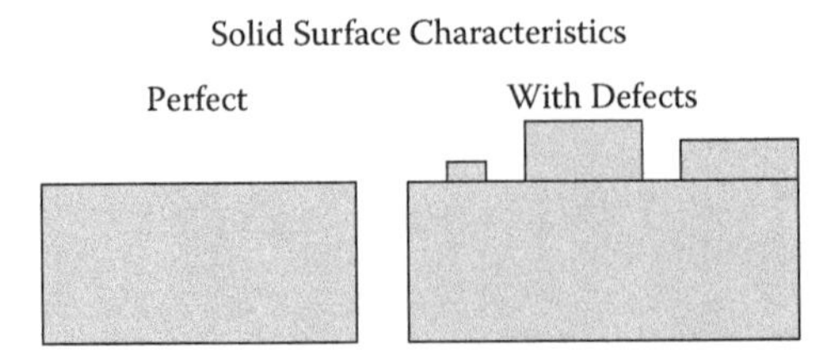

FIGURE 5.1 Solid surface molecule defects: (a) perfect crystal, (b) surface with defects.

larger than in its solid state. It is thus desirable at this stage to consider some of the basic properties of liquid–solid interfaces. The surface tension of a liquid becomes important when it comes in contact with a solid surface. The interfacial forces are responsible for self-assembly formation and stability on solid surfaces. The interfacial forces that are present between a liquid and solid can be estimated by studying the shape of a drop of liquid placed on any smooth solid surface (Figure 5.2). The balance of forces as indicated (using geometrical considerations), again, were analyzed very extensively in the 1800s by Young (1805), who related the different forces at the solid liquid boundary, and the contact angle, θ, as follows (Chattoraj and Birdi, 1984; Adamson and Gast, 1997; Birdi, 1997, 2002):

$$\text{Surface tension of solid } (\gamma_S) = \text{Surface tension of solid/liquid } (\gamma_{SL}) + \text{Surface tension of liquid } (\gamma_L) \cos(\theta) \quad (5.1)$$

$$\gamma_S = \gamma_L \cos(\theta) + \gamma_{SL} \quad (5.2)$$

$$\gamma_L \cos(\theta) = \gamma_S - \gamma_{SL} \quad (5.3)$$

where γ is the interfacial tension at the various boundaries between solid (S), liquid (L), and air (or vapor) phases. The relation of Young's equation is easy to understand as it follows from simple physics laws. At the equilibrium contact angle, all the relevant surface forces come to a stable state (Figure 5.2).

The equilibrium of forces is valid in all kinds of solid–liquid systems. The geometrical force balance is considered only in the X-Y plane. This assumes that the liquid does not affect the solid surface (in any physical sense). This assumption is

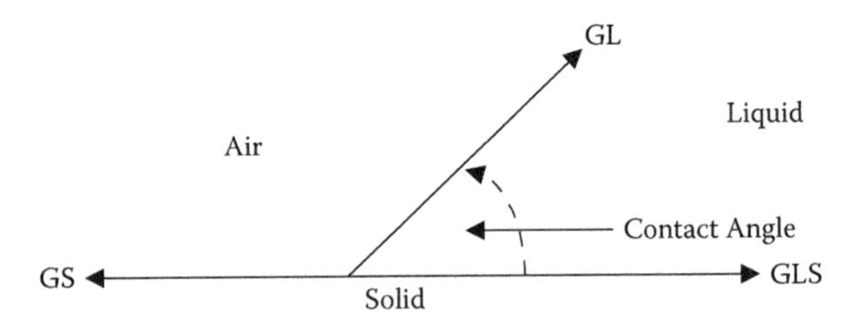

FIGURE 5.2 The surface force equilibrium between surface tensions of liquid (GL)–solid (GS)–liquid-solid (GLS)–contact angle (CA).

safe in most cases. However, only in very special cases, if the solid surface is soft (such as a contact lens), then one will expect that tangential forces will also need to be included in this equation (as extensively described in literature). Extensive data exists that convincingly supports the relation in Equation (5.2) for liquids and solids.

5.2 SOLID SURFACE TENSION: WETTING PROPERTIES OF SOLID SURFACES

One may ask, how can there be surface tension of a solid and how can one measure this force? These questions can be answered using the following arguments. Wetting of solid surfaces is well known when considering the difference between Teflon and metal surfaces. To understand the degree of *wetting*, between the liquid, L, and the solid, S, it is convenient to rewrite Equation (5.3) as follows:

$$\cos(\theta) = (\gamma_S - \gamma_{LS})/\gamma_L \tag{5.4}$$

which would then allow one to analyze the variation of γ with the change in the other terms. The latter is important because complete wetting occurs when there is no finite contact angle, and thus, $\gamma_L <> \gamma_S - \gamma_{LS}$. However, when $\gamma_L > \gamma_S - \gamma_{LS}$, then $\cos(\theta) < 1$, and a finite contact angle is present. The latter is the case when water, for instance, is placed on hydrophobic solid, such as Teflon, polyethylene, or paraffin. The addition of surfactants to water reduces γ_L, therefore, θ will decrease on introduction of such surface active substances. A complete discussion of the wetting of solids is beyond the scope of this book, and the reader is therefore encouraged to look up other standard textbooks on surface chemistry (Chattoraj and Birdi, 1984; Adamson and Gast, 1997; Birdi, 1997, 2002). The state of a fluid drop under dynamic conditions, such as *evaporation*, becomes more complicated (Birdi et al., 1988). However, in this text we are interested in the spreading behavior of when a drop of one liquid is placed on the surface of another liquid, especially when the two liquids are immiscible. The spreading phenomena by introducing a quantity, a spreading coefficient ($S_{a/b}$), is defined as (Harkins 1952; Adamson and Gast, 1997; Birdi, 2002):

$$S_{a/b} = \gamma_a - (\gamma_b + \gamma_{ab}) \tag{5.5}$$

where $S_{a/b}$ is the spreading coefficient for liquid b on liquid a, γ_a and γ_b are the respective surface tensions, and γ_{ab} is the interfacial tension between the two liquids. If the value of $S_{b/a}$ is positive, spreading will take place spontaneously; whereas if it is negative, liquid b will rest as a lens on liquid a.

However, the value of γ_{ab} needs to be considered as the equilibrium value, and therefore if the system is considered at nonequilibrium, then the spreading coefficients would be different. For example, the instantaneous spreading of benzene is observed to give a value of $S_{a/b}$ as 8.9 dyn/cm, and therefore benzene spreads on water. On the other hand, as the water becomes saturated with time, the value of (water) decreases and benzene drops tend to form lenses. The short-chain hydrocarbons such as hexane and hexene also have positive initial spreading coefficients

TABLE 5.1
Calculation of Spreading Coefficients, $S_{a/b}$, for Air–Water Interfaces (20°C)

Oil	$\gamma_{w/a} - \gamma o/a - \gamma o/w = S_{a/b}$	Conclusion
n-C16H34	72.8 – 30.0 – 52.1 = –0.3	Will not spread
n-octane	72.8 – 21.8 – 50.8 = + 0.2	Will just spread
n-octanol	72.8 – 27.5 – 8.5 = +36.8	Will spread

Notes: a, air; w, water; o, oil.

and spread to give thicker films. Longer-chain alkanes on the other hand, do not spread on water, for example, the $S_{a/b}$ for C16H34 (hexadecane)/water is –1.3 dyn/cm at 25°C. It is also obvious that since impurities can have very drastic effects on the interfacial tensions in Equation (2.24), the value of $S_{a/b}$ would be expected to vary accordingly (see Table 5.1).

The spreading of a solid substance, for example, cetyl alcohol (C_{18} H_{38} OH), on the surface of water has been investigated in some detail (Gaines, 1966; Adamson and Gast, 1997; Birdi, 2002). Generally, however, the detachment of molecules of the amphiphile into the surface film occurs only at the periphery of the crystal in contact with the air–water surface. In this system, the diffusion of amphiphile through the bulk water phase is expected to be negligible, because the energy barrier now includes not only the formation of a hole in the solid but also the immersion of the hydrocarbon chain in the water. It is also obvious that the diffusion through the bulk liquid is a rather slow process. Furthermore, the value of $S_{a/b}$ would be very sensitive to such impurities with regard to the spreading of one liquid upon another.

Another example is that the addition of surfactants (detergents) to a fluid dramatically affects its wetting and spreading properties. Thus, many technologies utilize surfactants for control of wetting properties (Birdi, 1997). The ability of surfactant molecules to control wetting arises from their *self-assembly* at the liquid–vapor, liquid–liquid, solid–liquid, and solid–air interfaces, and the resulting changes in the interfacial energies (Birdi, 1997). These interfacial self-assemblies exhibit rich structural detail and variation. In the case of oil spills on the seas, these considerations become very important. The treatment of such pollutant systems requires knowledge of the state of the oil. The thickness of the oil layer will be dependent on the spreading characteristics. The effect on ecology (such as birds and plants) will depend on the spreading characteristics. Young's equation at $liquid_1$–solid–$liquid_2$ has been investigated for various systems. This is found in such systems where the $liquid_1$–solid–$liquid_2$ surface tensions meet at a given contact angle. For example, the contact angle of a water drop on Teflon is 50° in octane (Chattoraj and Birdi, 1984) (Figure 5.3).

In this system the contact angle, θ, is related to the different surface tensions as follows:

$$\gamma_{\text{s-octane}} = \gamma_{\text{water-s}} + \gamma_{\text{octane-water}} \cos(\theta) \quad (5.6)$$

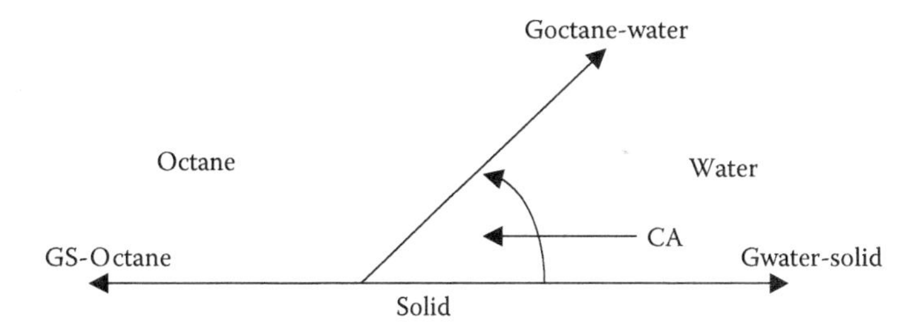

FIGURE 5.3 Contact angle at water–Teflon–octane interface. (See Equation 5.6.)

or

$$\cos(\theta) = (\gamma_{s\text{-octane}} - \gamma_{water\text{-}s})/\gamma_{octane\text{-}water} \tag{5.7}$$

This gives the value of $\theta = 50°$ when using the measured values of $(\gamma_{s\text{-octane}} - \gamma_{water\text{-}s})/\gamma_{octane\text{-}water}$. The experimental θ is the same. This analysis showed that the assumptions made in derivation of Young's equation are valid.

The most important property of a surface (solid or liquid) is its capability of interacting with other materials (gases, liquids, or solids). All interactions in nature are governed by different kinds of molecular forces (such as van der Waals, electrostatic, hydrogen bonds, dipole–dipole interactions). Based on various molecular models, the surface tension, γ_{12}, between two phases with γ_1 and γ_2, was given as (Chattoraj and Birdi, 1984; Adamson and Gast, 1997):

$$\gamma_{12} = \gamma_1 + \gamma_2 - 2\,\Phi_{12}\,(\gamma_1\,\gamma_2)^2 \tag{5.8}$$

where Φ_{12} is related to the interaction forces across the interface. The latter parameter will depend on the molecular structures of the two phases. In the case of systems such as alkane (or paraffin)–water, Φ_{12} is a unity, since the alkane molecule exhibits no hydrogen bonding property, while water molecules are strongly hydrogen bonded. Hence, all liquids and solids will exhibit γ of different kinds:

$$\text{Liquid surface tension: } \gamma_L = \gamma_{L,D} + \gamma_{L,P} \tag{5.9}$$

$$\text{Solid surface tension: } \gamma_S = \gamma_{S,D} + \gamma_{S,P} \tag{5.10}$$

This means that γ_S for Teflon arises only from *dispersion* (γ_{SD}) forces. On the other hand, a glass surface shows γ_S that will be composed of both $\gamma_{S,D}$ and $\gamma_{S,P}$. Hence, the main difference between Teflon and glass surface will arise from the $\gamma_{S,P}$ component of glass. This criteria has been of great importance when it involves the application of adhesives. The adhesive used for glass will need to bind to solid with both polar and apolar forces.

The values of γ_{SD} for different solids as determined from these analyses are given next.

Solid	γ_S	γ_{SD}	γ_{SP}
Teflon	19	19	0
Polypropylene	28	28	0
Polycarbonate	34	28	6
Nylon 6	41	35	6
Polystyrene	35	34	1
PVC	41	39	2
Kevlar 49	39	25	14
Graphite	44	43	1

In the case of polystyrene surfaces, it was found that the value of SP increased with treatment of sulfuric acid (due to the formation of sulfonic groups in the surface) (Birdi, 1997). This gave rise to increased adhesion of bacteria cells to the surfaces. Most of the notions and the physical laws of surfaces have been obtained by the studies of liquid–gas or $liquid_1$–$liquid_2$ interfaces. The solid surfaces have been studied in more detail, but this has taken place more recently. The *asymmetrical* forces acting at surfaces of liquids are much shorter than those expected on solid surface. This is due to the high energies that stabilize the solid structures. Therefore, when one considers solid surface then the *surface roughness* will need to be considered.

5.2.1 Definition of Solid Surface Tension (γ_{solid})

It was described earlier that the molecules at the surface of a liquid are under tension due to asymmetrical forces, which gives rise to surface tension. However, in the case of solid surfaces, one may not envision this kind of asymmetry as clearly. Although a simple observation might help one to realize that such surface tension analogy exists.

Let us compare the two systems:

- Water–Teflon
- Water–glass

For instance, let us analyze the state of a drop of water (10 μL) as placed on two different smooth solid surfaces, Teflon and glass. As seen, the contact angles are different (Figure 5.4).

Since the surface tension of water is the same in the two systems, then the difference in contact angle can only arise due to the *surface tension of solids* being different.

The surface tension of liquids can be measured directly (as described in Chapter 2). However, this is not possible in the case of solid surfaces. Experiments show that when a liquid drop is placed on a solid surface, the contact angle indicates that the molecules interact across the interface. Thus, this indicates that these data can be used to estimate the surface tension of solids.

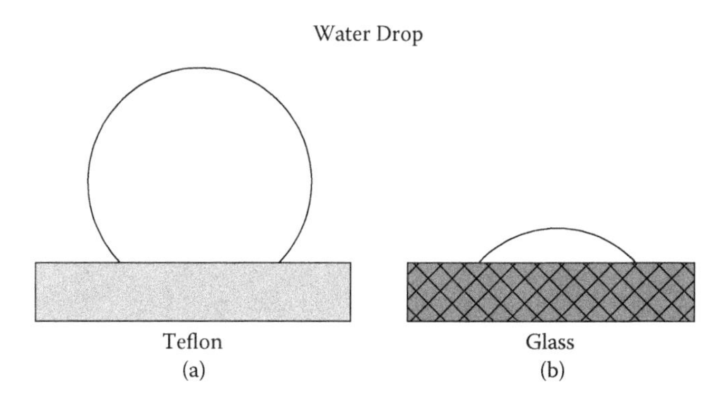

FIGURE 5.4 A drop of water on a smooth surface of (a) Teflon and (b) glass.

5.3 CONTACT ANGLE OF LIQUIDS ON SOLID SURFACES

As already mentioned, a solid in contact with a liquid leads to interactions related to the surfaces involved (that is, surface tensions of liquids and solids). The solid surface is being brought in contact with surface forces of the liquid (surface tension of liquid). If a small drop of water is placed on a smooth surface of Teflon or glass, Figure 5.4, it can be seen that these drops are different. The reason being is that there are three surface forces (tensions), which at equilibrium give rise to a contact angle. The relationship as given by Young's equation describes the interplay of forces (liquid surface tension, solid surface tension, liquid–solid surface tension) at the three-phase boundary line. This is regarded as if these forces interact along a line. Experimental data show that this is indeed true. The magnitude of contact angle is thus only dependent on the molecules nearest the interface, and independent of molecules much farther away from the contact line.

Further, one defines that:

- When the contact angle is less than 90°, the surface is wetting (such as water on glass).
- When the contact angle is greater than 90°, then the surface is nonwetting (such as water on Teflon).

Note that by treating the glass surface using suitable chemicals, the surface can be rendered hydrophobic. This is the same technology that is used in many utensils that are treated with Teflon or something similar.

5.4 MEASUREMENTS OF CONTACT ANGLES AT LIQUID–SOLID INTERFACES

The magnitude of the contact angle between a liquid and solid can be determined by various methods. The method to be used depends on the system as well as on the accuracy required. There are two common methods: (1) by direct microscope and a goniometer; and (2) by photography (digital analyses). It should be mentioned that the liquid drop that one generally uses in such measurements is very small, such as 10 to 100 μL. There are two different systems of interest: liquid–solid or $liquid_1$–solid–$liquid_2$. In the case of some industrial systems (such as oil recovery), the contact angle at high pressures and temperatures needs to be determined. In these systems, the value can be measured by using photography. Recently, digital photography has also been used, since these data can be analyzed by computer programs.

It is useful to consider some general conclusions from these data. A solid surface is defined as *wetting* if the contact angle is less than 90. However, a solid surface is designated as *nonwetting* if the contact angle is greater than 90. This is a practical and semiquantitative procedure. It has also been seen that water, due to its hydrogen bonding properties, exhibits a large contact angle on nonpolar surfaces (polyethylene, Teflon, etc.). On the other hand, lower contact angle values are found on polar surfaces (glass, mica). Next, it is important to consider how charged solid surfaces will exhibit properties that are related to contact angles. Metal surfaces will exhibit varying degrees of charges at the surface. Biological cells will exhibit charges that will affect adhesion properties to solid surfaces. However, in some applications one may change the surface properties by chemical modifications to the surface. For instance, PS (polystyrene) has some weak polar groups at the surface. If the surface is treated with H_2SO_4, this forms sulfonic groups. This leads to contact angle values lower than 30 (depending on the time of contact between sulfuric acid and PS surface) (Birdi, 2009). This treatment (or similar ones) has been used in many applications where the solid surface is modified to achieve a specific property. Since only the surface layer (a few molecules deep) is modified, the solid properties bulk does not change. This analysis shows the significant role of studying the contact angle of surfaces in relation to the application characteristics. A more rigorous analysis of these systems has been provided in the literature.

The magnitude of the contact angle of water (for example) is found to vary depending on the nature of the solid surface (Table 5.2). The magnitude of the contact angle is almost 100 on a waxed surface of car paint. The industry strives to create such surfaces to give $\theta > 150$, the so-called superhydrophobic surfaces. The large contact

TABLE 5.2
Contact Angles, θ, of Water on Different Solid Surfaces (25°C)

Solid	θ
Teflon (PTE)	108
Paraffin wax	110
Polyethylene	95
Graphite	86
AgI	70
Polystyrene	65
Glass	30
Mica	10

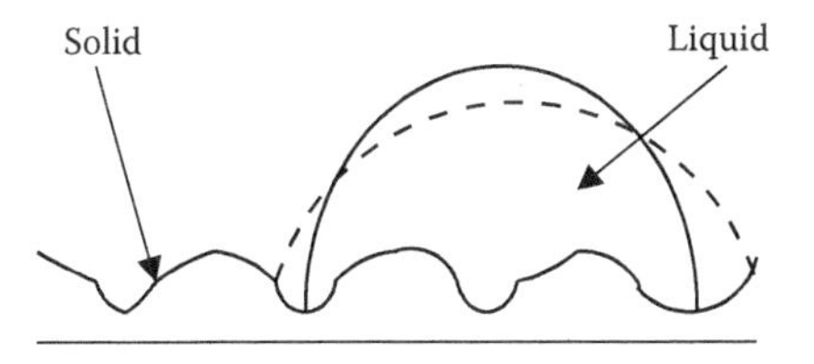

FIGURE 5.5 Analysis of a contact angle of a liquid drop on a rough solid surface.

angle means that water drops do not wet the car polish and are easily blown off by wind. Car polish is also designed to leave a highly smooth surface.

In many industrial applications, one is both concerned with smooth and rough surfaces. The analyses of contact angle on rough surfaces will be somewhat complicated than on smooth surfaces. The liquid drop on a rough surface, Figure 5.5, may show the real contact angle (solid line) or some lower value (apparent) (dotted line), dependent on the orientation of the drop.

However, no matter how rough the surface, the forces will be the same as those that exist between a solid and liquid. In other words, at a microscale, the balance of forces at the liquid–solid and contact angle, surface roughness has no effect. The surface roughness may show contact angle *hysteresis* if one makes the drop move, but this will arise from other parameters (e.g., wetting and dewetting). A *fractal* approach has been used to achieve a better understanding (Feder, 1988; Birdi, 1993; Koch, 1993).

In spite of its simple basis, Young's equation has been found to give useful analyses in a variety of systems. A typical data of cos(θ) various liquids on Teflon gave an almost straight-line plot (Figure 5.6).

The data can be analyzed by the following relation:

$$\cos(\theta) = k_1 - k_2\, \gamma_L \tag{5.11}$$

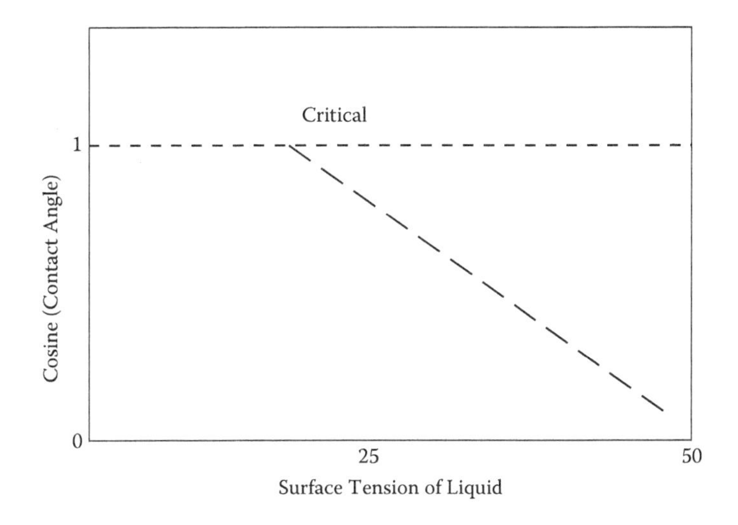

FIGURE 5.6 Plot of the contact angle cos(θ) of different liquids on Teflon.

TABLE 5.3
Some Typical γ_{cr} Values for Solid Surfaces (Estimated from Zisman Plots)

Surface Group	γ_{cr}
$-CF_2-$	18
$-CH_2-CH_3-$	22
Phenyl	30
Alkyl Chlorine	35
Alkyl hydroxyl	40

Source: Birdi, K. S., Editor, *Handbook of Surface & Colloid Chemistry—CD Rom*, 2nd ed., CRC Press, Boca Raton, FL, 2002.

This can also be rewritten as:

$$\cos(\theta) = 1 - k_3 (\gamma_L - \gamma_{cr}) \tag{5.12}$$

where γ_{cr} is the critical value of γ_L at cos(θ) equal to 0. The values of γ_{cr} have been reported for different solids using this procedure (as suggested by Zisman). The magnitude of γ_{cr} for Teflon of 18 mN/m thus suggests that -CF2- groups exhibit this low surface tension. The value of γ_{cr} for -CH2-CH3- alkyl chains gave a higher value of 22 mN/m than for Teflon. Indeed, from experience it has also been found that Teflon is a better water-repellant surface than any other material. The magnitudes of γ_{cr} for different surfaces is seen to provide much useful information (Table 5.3).

These data show that the molecular groups of different molecules determine the surface characteristics as related to γ_{cr}. In many cases the surface of a solid may not behave as desired, and therefore the surface is treated accordingly, which results in a change of the contact angle of fluids. For instance, the low surface energy polymers (polyethylene, PE) are found to change when treated with a flame or corona (as shown next).

Material	Liquid	Contact Angle
PE	Water	87
	Corona	55
PE (corona)	Water	66
	Corona	49

5.5 ADSORPTION OF GASES ON SOLID SURFACES

The most important solid surface property is its interaction with gases or liquids. The adsorption of a gas on a solid surface has been known to be of much importance in various systems (especially in industry involved with catalysis). The gas–solid surface phenomena can be analyzed as follows.

The molecules in gas are moving very fast, but on adsorption (gas molecules are more or less fixed) there will be a large decrease in kinetic energy (thus a decrease in entropy, ΔS). Movement of gas molecules in gas phase has much larger distances than when adsorbed on a solid surface.

Adsorption takes place spontaneously, which means:

$$\Delta G_{ad} = \Delta H_{ad} - T\Delta S_{ad}$$

that ΔG_{ad} is negative; this suggests that ΔH_{ad} is negative (exothermic). The adsorption of gas can be of different types. The gas molecule may adsorb as a kind of *condensation* process; it may under other circumstances react with the solid surface (chemical adsorption or chemisorption). In the case of chemoadsorption, a chemical bond formation is almost expected. On carbon while oxygen adsorbs (or chemisorb), one can desorb CO or CO_2. The experimental data can provide information on the type of adsorption. On porous solid surfaces, the adsorption may give rise to *capillary condensation*. This indicates that porous solid surfaces will exhibit some specific properties. The most used adsorption process in industry is in the case of catalytic reactions (for example, formation of $N\text{-}H_3$ from N_2 and H_2). It is thus apparent that in gas recovery from shale, the desorption of gas (mainly methane, CH_4) will be determined by the surface forces. The surface of a solid may differ in many ways from its bulk composition. Especially, such solids as commercial carbon black might contain minor amounts of impurities (such as aromatics, phenol, and carboxylic acid). This would render surface adsorption characteristics different than on pure carbon. It is therefore essential that in industrial production one maintains quality control of the surface from different production batches. Otherwise, the surface properties will affect the application. Another example arises from the behavior of glass powder and its adsorption characteristics for proteins. It has been found that if glass powder is left exposed to air, then its surface may become covered by pollutants from the air. This leads to a lower adsorption of proteins than on a clean surface. A silica surface has been considered to exist as O-Si-O as well as hydroxyl groups formed with water molecules. The orientation of the different groups may also be different at surface. Carbon black has been reported to possess different kinds of chemical groups on its surfaces. These different groups are aromatics, phenol, and carboxylic. These different sites can be estimated by comparing the adsorption characteristics of different adsorbents (such as hexane and toluene). When any clean solid surface is exposed to a gas the latter may adsorb on the solid surface to varying degree (as found from experiments). It has been recognized for many decades that gas adsorption on solid surfaces does not stop at a monolayer state. Of course, more than one layer (multilayer) adsorption will take place only if the pressure is reasonably high. Experimental data show this when a volume of gas adsorbed, v_{gas}, is plotted against P_{gas} (Figure 5.7).

From experimental analyses, it has been found that five different kinds of adsorption states exist (Figure 5.7). These adsorption isotherms were classified based on extensive measurements of v_{gas} versus p_{gas} data.

Type I—These are obtained for Langmuir adsorption.

Type II—This is the most common type where multilayer surface adsorption is observed.

Type III—This is a somewhat special type with almost only multilayer formation, such as nitrogen adsorption on ice.

Type IV—If the solid surface is porous, then this is found similar to type II.

Type V—On porous solid surfaces type III.

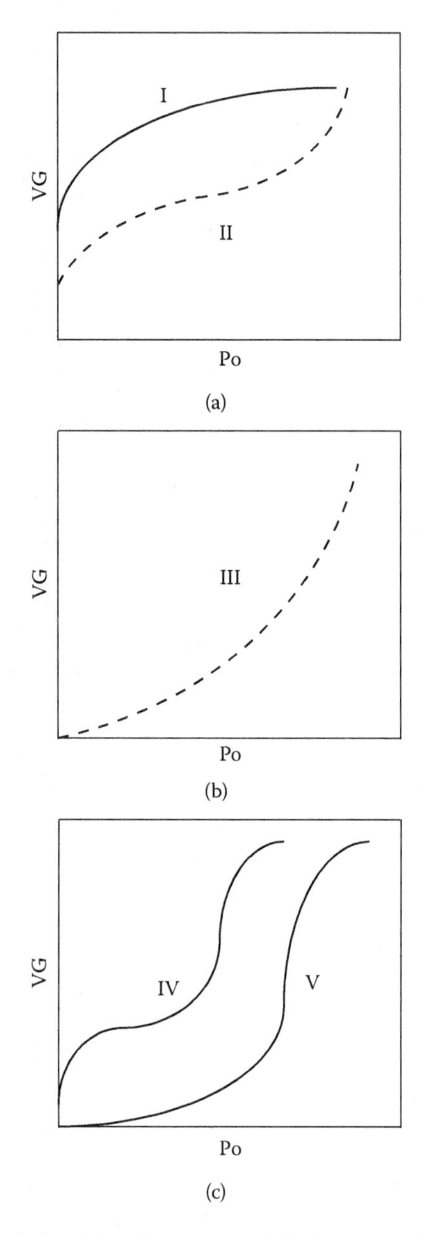

FIGURE 5.7 Plot of (relative volume) v_{gas} versus (relative pressure) P_{gas}. Types: (a) I and II, (b) III, (c) IV and V.

The pores in a porous solid surface are found to vary from 2 to 50 nm (*micropores*). *Macropores* are designated for larger than 50 nm. *Mesopores* are used for the 2 to 50 nm range.

5.5.1 Gas Adsorption on Solid Measurement Methods

5.5.1.1 Volumetric Change Methods

The change in the volume of gas during adsorption is measured directly in principle, and the apparatus is comparatively simple (Figure 5.8). One can use a mercury (other suitable liquid) reservoir beneath the manometer, and the burette is used to control the levels of mercury in the apparatus above. Calibration involves measuring the volumes of the gas (v_g) lines and of the void space (Figure 5.7). All pressure measurements are made with the right arm of the manometer set at a fixed zero point so that the volume of the gas lines does not change when the pressure changes. The apparatus, including the sample, is evacuated and the sample is heated to remove any previously adsorbed gas. A gas such as helium is usually used for the calibration, since it exhibits very low adsorption on the solid surface. After helium is pushed into the apparatus, a change in volume is used to calibrate the apparatus and the corresponding change in pressure is measured. A different gas (such as nitrogen) is normally used as the adsorbate if the surface area of a solid needs to be estimated. The gas is cooled by liquid nitrogen. The tap to the sample bulb is opened and the drop in pressure is determined. In the surface area calculations one uses a value of 0.162 nm^2 for the area of an adsorbed nitrogen molecule.

5.5.1.2 Gravimetric Gas Adsorption Methods

It is obvious that the amount of gas adsorbed on any solid surface will be of a very small magnitude. A modern sensitive microbalance is used to measure the adsorption isotherm. The sensitivity is very high since only the difference in weight change is

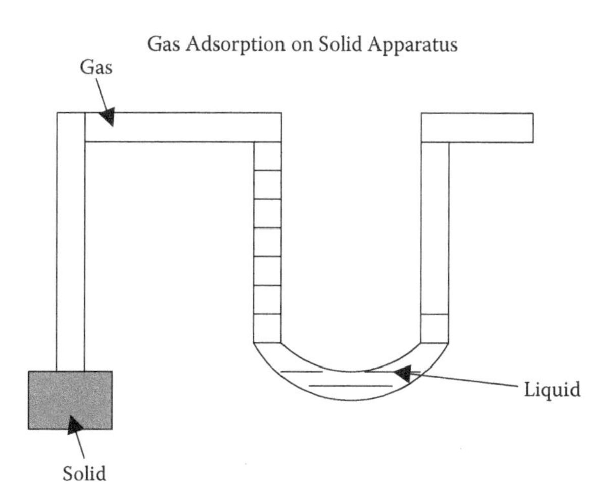

FIGURE 5.8 A typical gas adsorption on a solid surface apparatus.

measured. These microbalances can measure weight differences in the range of nanograms to milligrams. With such extreme sensitivity it is possible to measure the weight change caused by the adsorption of a single monolayer on a solid if the surface area is large. The normal procedure is to expose the sample to the adsorbate gas at a certain pressure, allowing sufficient time for equilibrium to be reached and then determining the mass change. This is repeated for a number of different pressures and the number of moles adsorbed as a function of pressure plotted to give an adsorption isotherm.

Microbalances (stainless steel) can be made to handle pressures as high as 120 Mpa (120 atm), since gases that adsorb weakly or boil at very low pressures can still be used.

5.5.1.3 Gas Adsorption on Solid Surfaces (Langmuir Theory)

A monolayer of gas adsorbs in the case where there are only a given number of adsorption sites for only a monolayer. This is the most simple adsorption model. The amount adsorbed, N_s, is related to the monolayer coverage, N_{sm}, as follows (see Appendix 5C for more details):

$$N_s/N_{sm} = a\,p/(1 + a\,p) \tag{5.13}$$

where p is the pressure and a is dependent on the energy of adsorption. This equation can be rearranged:

$$p/N_s = (1/(a\,N_{sm}) + p/N_{sm}) \tag{5.14}$$

From the experimental data, one can plot p/N_s versus p. The plot will be linear and the slope is equal to $1/N_{sm}$. The intersection gives the value of a. Charcoal is found to adsorb 15 mg of N_2 as monolayer. Another example is that of adsorption of N_2 on mica surface (at 90 K). The following data were found:

Pressure/Pa	Volume of Gas Adsorbed (at Standard Temperature and Pressure, STP)
0.3	12
0.5	17
1.0	24

In this equation one assumes:

- That the molecules adsorb on definite sites.
- That the adsorbed molecules are stable after adsorption.

The surface area of the solid can be estimated from the plot of p/N_s versus p. Most data fit this equation under normal conditions and is therefore widely applied to analyze adsorption processes.

Langmuir adsorption is found for the data of nitrogen on mica (at 90 K). The data were as follows:

p = 1/Pa   2/Pa
Vs = 24 mm^3 28 mm^3

This shows that the amount of gas adsorbed increases by a factor of 28/24 = 1.2 when the gas pressure increases twofold.

5.5.1.4 Various Gas Adsorption Equations

The gas adsorption on sold surface data has been analyzed by different models. Other isotherm equations begin as an alternative approach to the developed equation of state for a two-dimensional ideal gas. The assumptions in deriving the equation of state form the isotherm equation. As mentioned earlier, the ideal equation of state is as follows:

$$\Pi\, A = k_B T$$

In combination with the Langmuir equation one can derive the following relation between N_s and p:

$$N_s = K\, p \tag{5.15}$$

where K is a constant. This is the well-known Henry's law relation and it is valid for most isotherms at low relative pressures. In these situations where the ideal Equation 5.15 does not fit the data, the van der Waals equation type of corrections have been suggested.

The *adsorption–desorption process* is of interest in many systems (such as cement). The water vapor may condense in the pores after adsorption under certain conditions. This may be studied by analyzing the adsorption–desorption data (Figure 5.9).

As mentioned elsewhere, the fracking process for gas recovery from shale deposits is a process where adsorbed gas is released by the process. In some systems, adsorption of gas molecules proceeds to higher levels where multilayers are

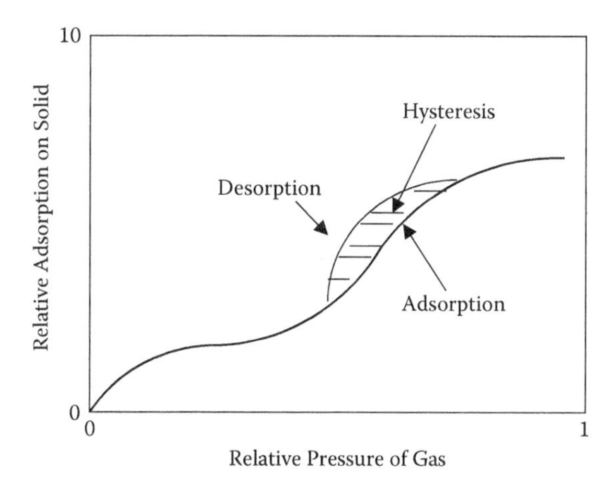

FIGURE 5.9 Adsorption (Ns/Nsm = relative adsorption) versus pressure (*p/po* = relative pressure) of a gas on a solid.

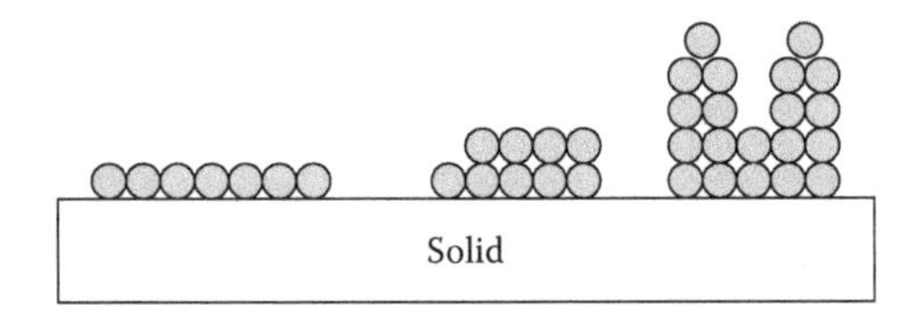

FIGURE 5.10 BET model for multilayer adsorption on solids.

observed. These are found from data analyses when multilayer adsorption takes place (Figure 5.10).

The Brunauer–Emmett–Teller (BET) equation has been derived for multilayer adsorption data.

The enthalpy involved in multilayers is related to the differences and was defined by BET theory as:

$$E_{BET} = \exp[(E_l - E_v)/RT] \tag{5.16}$$

where E_l and E_v are enthalpies of desorption. The BET equation thus after modification of the Langmuir equation becomes:

$$p/(N_s (p^o - p)) = 1/E_{BET}\, N_{sm} + [(E_{BET} - 1)/(E_{BET}\, N_{sm})\ (p/po)] \tag{5.17}$$

From a plot of adsorption data on the left-hand side of this equation versus relative pressure (p/po) allows one to estimate N_{sm} and E_{BET}. The magnitude of E_{BET} is found to give either data plots that are as type III or II. If the value of E_{BET} is low, which means the interaction between adsorbate and solid are weak, then type III plots are observed. This has been explained as arising from the fact that if $E1 \triangleq Ev$, then molecules will tend to form multilayers in patches rather than adsorb on a naked surface. If a strong interaction exists between the gas molecule and the solid, $E_l > E_v$, then type II plots are observed. The monolayer coverage is clearly observed at low values of p/po.

5.6 ADSORPTION OF SUBSTANCES FROM SOLUTIONS ON SOLID SURFACES

Any clean solid surface is actually an active center for adsorption from the surroundings, for example, air or liquid. A perfectly cleaned metal surface, when exposed to air, will adsorb a single layer of oxygen or nitrogen (or water) (degree of adsorption will depend on the system). The most common example of much importance is the process of corrosion (an extensive economic cost) of iron when exposed to air. When a completely dry glass surface is exposed to air (with some moisture) the surface will adsorb a monolayer of water (Figure 5.11). In other words, the solid surface is not as inert as it may seem to the naked eye. This has many consequences in industry, such as corrosion control.

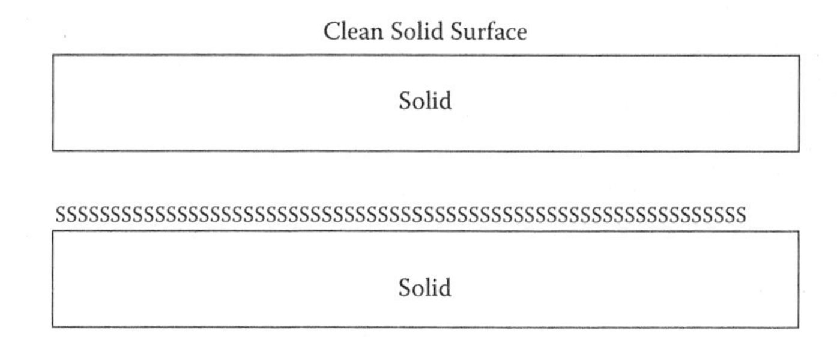

FIGURE 5.11 A perfectly clean surface on a solid exposed to air or any other gas. Adsorbed molecules (shown as S) will have a specific geometrical arrangement.

Thermodynamics of Adsorption

Activated charcoal or carbon (with surface area of over 1000 m^2 per gram) is widely used for vapor adsorption and in the removal of organic solutes from water. These materials are used in industrial processes to purify drinking water and swimming pool water, to decolorize sugar solutions as well as other foods, and to extract organic solvents (especially trace amounts of dangerous substances). They are also used as the first, oral treatment in hospitals for cases of poisoning. Activated charcoal can be made by heat degradation and partial oxidation of almost any carbonaceous matter of animal, vegetable, or mineral origin. For convenience and economic reasons it is usually produced from bones, wood, lignite, or coconut shells. The complex three-dimensional structure of these materials is determined by their carbon-based polymers (such as cellulose and lignin) and it is this backbone that gives the final carbon structure after thermal degradation. These materials, therefore, produce a very porous high-surface-area carbon in solid form. In addition to a high surface area, the carbon has to be "activated" so that it will interact with and physisorb (i.e., adsorb physically without forming a chemical bond) a wide range of compounds. This activation process involves controlled oxidation of the surface to produce polar sites.

Adsorption of Acetic Acid onto Activated Charcoal

Adsorption at liquid surfaces can be monitored using the Gibbs adsorption isotherm since the surface energy, γ, of a solution can be readily measured. However, for solid substrates this is not the case and the adsorption density has to be measured in some other manner. In the present case the concentration of adsorbate in solution will be monitored. In place of the Gibbs equation we can use a simple adsorption model based on the mass action approach.

Solid surfaces will have molecules arranged at the surface in a very well-defined geometrical arrangement. This will give rise to surface forces that will determine the adsorption of a particular substance. On any solid surface, a certain number of possible adsorption sites per gram (N_m) are to be expected. This is the number of sites where any adsorbate can freely adsorb. There will be a fraction contact angle filled by one adsorbing solute. An adsorption–desorption process will exist at the surface, which is also expected. The rate of adsorption will be given as:

$$\text{(Concentration of solute)}(1 - \theta)\, N_m \tag{5.18}$$

and the rate of desorption will be given as:

$$\text{(Concentration of solute)}(\theta)\, N_m \tag{5.19}$$

It is known that at equilibrium these rates must be equal:

$$k_{ads}\, C_{bulk}\, (1 - \theta)\, N_m = k_{des}\, \theta\, N_m \tag{5.20}$$

where k_{ads} and k_{des} are the respective proportionality constants, and C_{bulk} is the bulk solution concentration of solute. The equilibrium constant, $K_{eq} = k_{ads}/k_{des}$, which gives:

$$C_{bulk}/\theta = C_{bulk} + 1/K_{eq} \tag{5.21}$$

and since $\theta = N/N_m$, where N is the number of solute molecules adsorbed per gram of solid, one can write:

$$C_{bulk}/N = C_{bulk}/N_m + 1/(K_{eq}\, N_m) \tag{5.22}$$

Thus, measurement of N for a range of concentrations (C) should give a linear plot of C_{bulk}/N against C_{bulk}, where the slope gives the value of N_m and the intercept the value of the equilibrium constant K_{eq}. This model of adsorption was suggested by Irving Langmuir and is referred to as the *Langmuir adsorption isotherm*. The aim of this experiment is to test the validity of this isotherm equation and to measure the surface area per gram of charcoal, which can easily be obtained from the measured N_m value, if the area per solute molecule is known.

Adsorption experiments are carried out as follows. The solid sample (for example, activated charcoal) is shaken in contact with a solution with known concentration of acetic acid. After equilibrium is reached (approximately after 24 hours) the amount of acetic adsorbed is determined. One can determine the concentration of acetic acid by titration with an NaOH solution.

One may also use dye solutions (such as methylene blue), and after adsorption the amount of dye in a solution is measured by any convenient spectroscopic method (VIS or UV or fluorescence spectroscopy).

5.6.1 Solid Surface Area Determination

As far as surface chemistry is concerned, a solid particle is a very important substance with regard to its surface characteristics. In all applications where finely divided powders are used (such as talcum, cement, charcoal powder) the property of these will depend mainly on the surface area per gram (varying from a few square meters [talcum] to over 1000 m^2/g [charcoal]). For example, if charcoal is used to remove some chemical (such as coloring substances or other pollutants) from wastewater, then it is necessary to know the amount of absorbent needed to fulfill the process. In other words, if a 1000 m^2 area is needed for adsorption when using charcoal, then 1 g of solid will be required. In fact, under normal conditions, swallowing charcoal is considered dangerous, because it will lead to the removal of essential substances (such as lipids and proteins) from the stomach lining. The estimation of surface area of finely divided solid particles from solution adsorption studies is subject to many of the same considerations as in the case of gas adsorption but with the added complication that larger molecules are involved whose surface orientation and pore penetrability may be uncertain. A first condition is that a definite adsorption model is obeyed, which in practice means that area determination data are valid within the simple Langmuir equation (Equation 5.22) relation. The constant rate is found, for example, from a plot of the data according to Equation (5.22), and the specific surface area then follows from Equation 5.20 and Equation 5.21. The surface area of the adsorbent is generally found easily from the literature. In the case of gas adsorption where the BET method is used, it is reasonable to use the van der Waals area of the adsorbate molecule: moreover, being small or even monoatomic, surface orientation is not a major problem. In the case of adsorption from solutions, however, the adsorption may be chemisorption. In the literature, fatty acid adsorption for surface area estimation has also been used. This is useful since fatty acids are known to pack perpendicular to the surface (self-assembly monolayer formation) and with the close-packed area per molecule of 20.5 Å^2. This seems to be true for adsorption on such diverse solids as carbon black and for TiO_2. In all of these cases, the adsorption is probably chemisorption in involving hydrogen bonding or actual salt formation with surface oxygen. If polar solvents are used to avoid multilayer formation on top of the first layer, the apparent area obtained may vary with the solvent used. In the case of stearic acid on a graphitized carbon surface, Graphon, the adsorption, while obeying the Langmuir equation, appears to be physical, with the molecules flat on the surface.

As another example, the adsorption of surfactants on polycarbonate indicated that depending on the surfactant and concentration the adsorbed molecules might be lying flat on the surface perpendicular to it or might form a bilayer.

A second class of adsorbates of which much use has been made is that of using dyes. This method is appealing because of the ease with which analysis may be made colorimetrically. The adsorption generally follows the Langmuir equation. An apparent molecular area of 19.7 Å^2 for methane blue on Graphon was found (the actual molecular area is 17.5 Å^2). The fatty acid adsorption method has been used by many investigators. One has used pyridine adsorption on various oxide obtain surface areas. The adsorption data followed the Langmuir equation, the effective molecular area of pyridine is about 24 Å^2 per molecule. In literature many different approaches have been proposed to estimate the surface area of a solid. Surface areas may be estimated from the exclusion of like charged ions from a charged interface. This method is intriguing in that no estimation of either site or molecular area is needed. In general, however, surface area determination by means of solution adsorption studies, while convenient experimentally, may not provide the most correct information. Nonetheless, if a solution adsorption procedure has been standardized for a given system, by means of independent checks, it can be very useful determining relative areas of a series of similar materials. In all cases, it is also more real as it is what happens in real life.

Adsorption experimental method for solid surfaces: The typical procedure used was 1.0 g of alumina powder (as an example) and 10 mL of solution of detergent with varying concentration. The mixture was shaken and the concentration of detergent was estimated by some suitable method. It was found that equilibrium was obtained after 2 to 4 hours.

A detergent, such as dodecylammonium chloride, was found to adsorb 0.433 mM per g of alumina with a surface area of 55 m^2/g. The surface area of alumina as determined from stearic acid adsorption (and using the area/molecule of 21 $Å^2$ from monolayer), gave this value of 55 m^2/g.

These data can be analyzed in more detail.

$$\text{Surface area of alumina} = 55\ m^2/g$$

$$\text{Amount adsorbed} = 0.433\ mM/g$$

$$= 0.433\ 10^{-3}\ M\ 6\ 10^{23}\ \text{molecules}$$

$$= 0.433\ 10^{20}\ \text{molecules}$$

$$\text{Area/molecule} = 55\ 10^{4}\ cm^2\ 10^{16}\ Å^2/0.433\ 10^{20}\ \text{molecule}$$

$$= 55/0.433\ Å^2$$

$$= 127\ Å^2$$

The adsorption isotherms obtained for various detergents showed a characteristic feature that a equilibrium value was obtained when the concentration of detergent was over critical micelle concentration (CMC). The adsorption was higher at 40°C than at 20°C. However, the shapes of the adsorption curves were the same (Birdi, 2002). Detergents adsorb on solids until CMC, after which no more adsorption is observed. This shows that after the solid surface is covered by a monolayer of detergent, the adsorption stops.

One can also calculate the amount of a small molecule, such as pyridine (mol. wt. 100), adsorbed as a monolayer on charcoal with 1000 m^2/g. In the following these data are delineated:

$$\text{Area per pyridine molecule} = 24\ Å^2 = 24\ 10^{-16}\ cm^2$$

$$\text{Surface area of 1 g charcoal} = 1000\ m^2 = 1000\ 10^{4}\ cm^2$$

$$\text{Molecules pyridine adsorbed} = 1000\ 10^4\ \text{cm}^2/24\ 10^{-16}\ \text{cm}^2/\text{molecule}$$
$$= 40\ 10^{20}\ \text{molecules}$$

$$\text{Amount of pyridine adsorbed/g of charcoal} = (40\ 10^{20}\ \text{molecules}/6\ 10^{23})100$$
$$= 0.7\ \text{g}$$

This is a useful example to illustrate the application of charcoal (or similar substances with large surface area per gram) in removal of contaminants by adsorption. In case of poison ingestion, one uses active charcoal to remove poison in stomach through adsorption. Talcum is effective by covering large areas of skin and providing protection, for instance, skin irritation.

5.6.2 Interaction of a Solid with Liquids (Heats of Adsorption)

A solid surface interacts with its surrounding molecules (in gas or liquid phase) with varying degree. For example, if a solid is immersed into a liquid the interaction between the two bodies will be of interest. The interaction of a substance with a solid surface can be studied by measuring the heat of adsorption (in addition to other methods). The information as to whether the process is exothermic (heat is produced) or endothermic (heat is absorbed) is needed. This leads to the understanding of the mechanism of adsorption and helps in application and design of the system. Calorimetric measurements have provided much useful information. When a solid is immersed in a liquid (Figure 5.12), in most cases there is a liberation of heat:

$$q_{imm} = E_S - E_{SL} \tag{5.23}$$

where E_S and E_{SL} are the surface energy of solid surface and the solid surface in liquid, respectively. The quantity q_{imm} is measured from calorimetry where temperature change is measured after a solid (in finely divided sate) is immersed in a given liquid. Since these measurements can be carried out with microcalorimeter sensitivity, many systematic and detailed data have been reported in the literature. When a polar

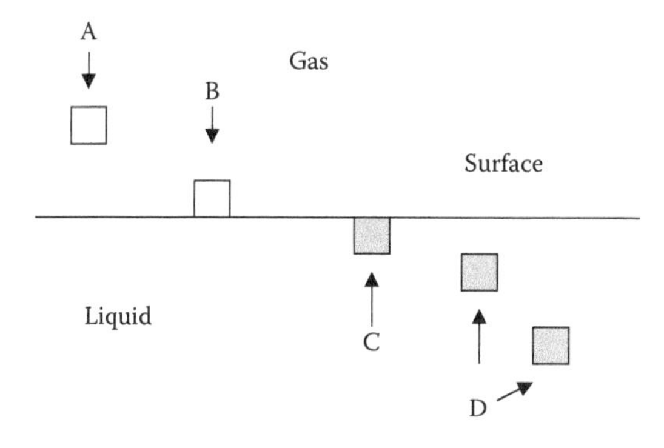

FIGURE 5.12 Solid immersion process in a liquid.

TABLE 5.4
Heats of Immersion (q_{imm}) (erg/cm^2 at 25°C) of Solids in Liquids

Solid Liquid	Polar (H2O-C2H5OH)	Liquid Nonpolar (C6H14)
Polar (TiO_2, Al_2O_3, glass)	400–600	100
Nonpolar (Graphon, Teflon)	6–30	50–100

solid surface is immersed in a polar liquid there will be a larger q_{imm} than if the liquid was an alkane (nonpolar). Values of some typical systems are given in Table 5.4. These data further show that such studies are sensitive to the surface purity of solids. For example, if the surface of glass powder is contaminated with nonpolar gas, then its q_{imm} value will be lower than in the case of pure glass surface.

5.7 SOLID SURFACE ROUGHNESS (DEGREE OF ROUGHNESS)

The nature of solid surface (surface area, surface roughness) plays an important role in many applications (e.g., cars, jewelry, television, computers, furniture, windmills, airplanes, ships). For example, a significant saving is observed in the fuel usage if the ship hull is smooth than if it is rough (e.g., due to algae). In many applications it is the main criteria. For instance, the friction decreases appreciably as the surface of a solid becomes smooth. This arises from the fact that the number of surface molecules that are able to come in contact with another solid or liquid phase are reduced (Figure 5.13). In another example, in the building of tunnels, there is a great reduction in the use of energy when the friction between the bore surface and the rock is reduced by special additives (such as soaps).

Thus in some cases one prefers roughness (high friction) (e.g., roads, shoe soles, tires), whereas in other systems (e.g., glass, office table) one requires smooth solid surface characteristics. The solid surfaces that are found are manufactured by different methods: sawn, cut, turned, polished, or chemically treated. All of these procedures

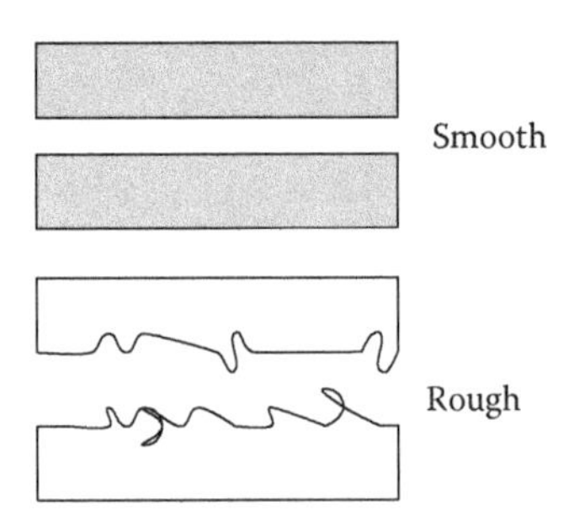

FIGURE 5.13 Profile of a solid rough and smooth surface.

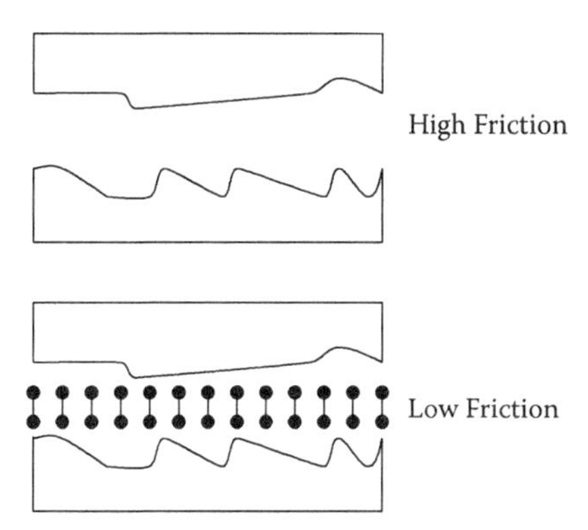

FIGURE 5.14 Friction reduction due to lubrication by adsorbed film (bilayer) of amphiphiles (as a lubricant) (schematic and highly idealized model).

leave the solid surface rough, with varying degree. In industry, one finds various methods that can characterize the roughness. Polishing is also an important application area of surface chemistry of solids. The surface layer produced after polishing may or may not remain stable after exposure to its surroundings (air, other gases, oxidation). The polishing industry is much dependent on the behavior of the surface molecular behavior.

Friction technology between $solid_1$–$solid_2$: Friction is defined as the resistance to sliding between two bodies. In everyday life friction is almost present all over and around us. In technology, such as when boring tunnels, friction plays a very important role. In fact, friction can be a very significant economical factor in such industrial processes. Thus, large savings can be made by reducing the degree of friction. Friction is lowered by using a suitable lubricant (specific for each system) (Lara et al., 1998; Biresaw and Mittal, 2008). Another example is the car industry, where lubrication of the different moving parts plays an important role in the working processes. In the case when solids are very close, then the surface roughness becomes the determining factor. This means that the resistance is higher between two rough surfaces. The degree of plasticity or deformation of the solids will also affect the friction. Further, a lubricant will have a high resistance of sliding if its viscosity is high. Thus, when one solid is sliding or rubbing against another there are a variety of parameters that need to be investigated. These parameters are also called as Tribology (related to rubbing). The *coefficient of friction* can be appreciably reduced if the boundary lubrication decreases the force field. This may be achieved by adsorbed films (Figure 5.14).

5.8 SURFACE TENSION OF SOLID POLYMERS

The surface tension of polymers (synthetic polymers, such as plastics; and biopolymers, such as proteins and gelatin) is indeed of much interest in many areas. In industry where plastics are used, the adhesion of these materials to other materials (such as steel and glass) is of much interest. The adhesion process is very complex since the demand on quality and control is very high. This is also due to the fact that many adhesion systems are part of life-sustaining processes (such as implants).

The forces involved in adhesion need to be considered. Let us consider some typical examples.

5.9 DIVERSE APPLICATIONS

5.9.1 Particle Flotation Technology of Solid Particles to Liquid Surfaces

In mineralogy, the technique of separating particles of minerals from water media are of great interest. Minerals or metals in pure form (such as gold) are only found in rare cases. The earth's surface consists of a variety of minerals (major components are iron, silica, oxides, calcium, magnesium, aluminum, chromium, cobalt, and titanium). Minerals, such as those found in nature are always mixed with different kinds (for example, zinc sulfide and feldspar minerals). In order to separate zinc sulfide, the mixture is suspended in water, which makes air bubbles achieve separation. This process is called *flotation* (ore, heavier than water, is floated by bubbles).

Flotation is a technical process in which suspended particles are clarified by allowing the suspended particles to float to the surface of the liquid medium (Fuerstenau et al., 1985). The material can thus be removed by skimming at the surface. This is economically much cheaper than any other process. If the suspended particles are heavier than the liquid (such as minerals) then one uses gas (air or CO_2 or other suitable gas) bubbles to enhance the flotation.

Froth flotation commences by grinding the rock, which is used to increase the surface area of the ore for subsequent processing and break the rocks into the desired mineral and gangue (which then has to be separated from the desired mineral); the ore is ground into a fine powder. The desired mineral is rendered hydrophobic by the addition of a surfactant or collector chemical; the particular chemical depends on the mineral being refined, for example, pine oil is used to extract copper. This slurry (more properly called the pulp) of hydrophobic mineral-bearing ore and hydrophilic gangue is then introduced to a water bath, which is aerated, creating bubbles. The hydrophobic grains of mineral-bearing ore escape the water by attaching to the air bubbles, which rises to the surface, forming foam (more properly called froth). The froth is removed and the concentrated mineral is further refined.

The flotation industry is a very important area in metallurgy and other related processes. The flotation method is based on treating a suspension of minerals (ranging in size from 10 to 50 μm) in the water phase to air (or some other gas) bubbles (Figure 5.15).

Flotation leads to separation of ores from the mixtures. Especially, in modern mineral industry where rare metals are being processed, this technique is being widely applied. It has been suggested that among other surface forces, the contact angle plays an important role. The gas (air or other gas) bubble as attached to the solid particle should have a large contact angle for separation.

Bubbles as needed for flotation are created by various methods. These may include air injection, electrolytic methods, and vacuum activation.

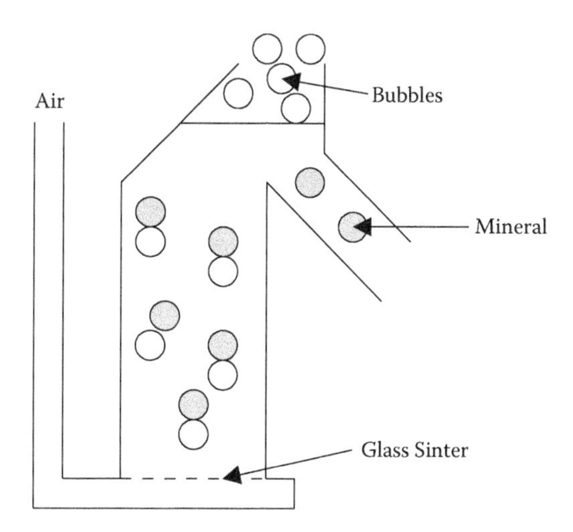

FIGURE 5.15 Flotation of mineral particles as aided by air bubbles.

In a laboratory experiment (Adamson and Gast, 1997) one may use the following recipe. To a 1% sodium bicarbonate solution one can add a few grams of sand. Then if some acetic acid (or vinegar) is added, the bubbles of CO_2 produced cling to the sand particles and thus make these float to the surface. It must be mentioned that in wastewater treatments the flotation method is one of the most important procedures. When rocks in crushed state are dispersed in water with suitable surfactants (also called collectors in industry) to give stable bubbles on aeration, then hydrophobic minerals will be floated to the surface by the attachment of bubbles, while the hydrophilic mineral particles will settle to the bottom. The preferential adsorption of the collector molecules on a mineral makes it hydrophobic. Xanthates have been used for the flotation of lead and copper. In these examples, it is the adsorption of xanthate ions on the mineral particles that dominates the flotation.

5.9.2 Polishing of Solid Surfaces

The reflection of light from a smooth solid surface determines the degree of polish. The industrial applications are very large, considering the impact of design and looks on both sales and user effects (e.g., cars, household appliances, furniture). A very sensitive procedure for the determination of the surface roughness has been to use atomic force microscopy (AFM) (Birdi, 2003).

5.9.3 Finely Divided Solid Particles (Powder Surface Technology)

It is a common observation that sand particles such as those that are found on a beach behave differently in comparison to talcum powder, with regard to their surfaces. This difference arises from the fact that any solid material when transformed into fine powder creates a new material (concerning its surface properties and applications). The technology related to this subject is very expansive (such as, talcum powder, cement, and clay industry).

The following areas of powder science are currently being investigated:

- Formation and synthesis of particles by different procedures (e.g., grinding, crushing, precipitation)
- Modification of particles by agglomeration, coating, and so forth
- Characterization of the size, shape, surface area, pore structure, and strength of particles
- Origins and effects of interparticle forces
- Packing, failure, flow, and permeability of assemblies of particle
- Particle–particle interactions and suspension rheology
- Handling and processing operations (e.g., slurry flow, fluidization, pneumatic conveying)
- Applications of particle technology in production of pharmaceuticals, chemicals, foods, pigments, structural, and functional materials, and in environmental and energy related areas.

The effect of synthesis parameters on precipitation of nanocrystalline boehmite from aluminate solutions was investigated recently. Nanocrystalline boehmite (AlOOH) is a useful and effective material for the production of Al_2O_3, as applied in many industrial applications as catalyst or catalyst support, membranes, and adsorbents. The preparation conditions applied in the production step of nanocrystalline boehmite strongly affects its morphology, which in turn is reflected in the final transition to alumina. In this work, a precipitation method for the production of nanocrystalline boehmite is described by studying the effects of pH, temperature, and aging time on the morphology of the final precipitate. The experiments were performed at 30°C, 60°C, and 90°C, under moderate pH (varying from pH 5 to 7) conditions and 1 week of aging in the mother liquor. It is noteworthy that in these experiments a unique starting solution is used and also the mixing procedure is unique. The starting solution used is a supersaturated sodium aluminate solution. On the other hand, the mixing procedure does not follow the normal route of addition of the neutralization agent (acid) to the aluminate solution. Amorphous boehmite was prepared at 30°C and pH 7 under prolonged aging conditions. At 60°C the formation of pure nanocrystalline boehmite with crystallites 3 to 8 nm was formed at pH 6 and pH 7 after aging in the mother liquor, while at the higher temperature of 90°C the formation of pure nanocrystalline boehmite with crystallite size between 3 and 13 nm was obtained at pH 5, pH 6, and pH 7. Aging and temperature influenced the crystallinity of the precipitated phases, with prolonged aging and high temperatures inducing high crystallinity. The pH conditions also had a strong effect on the crystallite size of precipitates. Actually, for the same temperature and aging time, the higher the pH, the larger the crystallites of the precipitates.

Nanocrystalline boehmite can be synthesized by neutralization of AlOOH liquor under atmospheric conditions (60°C/pH 6), proper aging, and a modification of the usual neutralization procedure.

At present there are many research reports related to the development and processing of inorganic particles ranging from the nanoscale of less than 10 nm to the microscale. The processing comprises various stages of materials fabrication from particles (powders), starting with the particle synthesis to the point of forming and densification to reach the final product.

Some of these projects are related to following subjects that deal with the development and functionalization of nanoparticles for the following:

- Drug and gene delivery
- Self-assembling of nanoparticles to achieve unique surface structures with controlled porosity and (bio)functionality
- The creation of three-dimensional building blocks for further processing, including sintered products

The main aim has been to understand at a fundamental level these various steps with a view to improving the processing routes for materials of technological importance and to contribute to the development of new and intelligent materials. The high quality of particles with regard to narrow size distribution, form factors, functions, and so forth, is guaranteed by the high level of characterization methods for particles in the nano and micro ranges.

Appendix 5C: Gas Adsorption on Solid Surfaces—Essential Principle Theory

From physicochemical principles it is known that any pure surface, especially a solid, means that it is surrounded by no other foreign molecule (which means it is under a vacuum). However, as soon as there is a foreign molecule in the gas phase, the latter will adsorb to some extent dependent on the physical conditions (temperature and pressure). On any solid surface, gas will adsorb or desorb under specific conditions (for example, in gas recovery from shale deposits). Gas molecules will adsorb and desorb at the solid surface as determined by different parameters. At equilibrium the rates of adsorption (R_{ads}) and desorption (R_{des}) will be equal. The surface can be described as consisting of different kinds of surfaces:

Total surface area = $A_t = A_o + A_m$
Area of clean surface = A_o
Area covered with gas = A_m
Enthalpy of adsorption = E_{ads} (energy required to adsorb a molecule from gas phase to the solid surface)

One can write the following relations:

$$R_{ads} = k_a p A_o \tag{5C.1}$$

$$R_{des} = k_b A_m \exp(-E_{ads}/RT) \tag{5C.2}$$

where k_a and k_b are constants.

At equilibrium:

$$\text{Rate of adsorption } (R_{ads}) = \text{Rate of desorption } (R_{des}) \tag{5C.3}$$

and the magnitude of A_o is a constant.

Further we have:

Amount of gas adsorbed = N_s
Monolayer capacity of the solid surface = N_{sm}

By combining these relations and

$$N_s/N_{sm} = A_m = A_t \quad (5C.4)$$

we get the well-known Langmuir adsorption equation (Birdi, 2008):

$$N_s = N_{sm}/(a\ p)/(1 + (a\ p)) \quad (5C.5)$$

Additionally, the heat of adsorption has been investigated. For example, the amount of Kr adsorbed on AgI increases when the temperature is decreased from 79 K (0.13 cc/g) to 77 K (0.16 cc/g). This data allows one to estimate the isosteric heat of adsorption (Jaycock and Parfitt, 1981; Birdi, 2008):

$$d\ (Ln\ P/dT) = q_{ads}/RT^2 \quad (5C.6)$$

The magnitude of q_{ads} was in the range of 10 to 20 kJ/mol.

6 Wetting, Adsorption, and Cleaning Processes

6.1 INTRODUCTION

In everyday life there are various systems where a liquid comes in contact with the surface of a solid. Contact angle studies of liquid–solid systems have shown that wetting is dependent on different parameters. It has been found that when a liquid comes in contact with a solid, there are a few specific processes that need to be analyzed. These processes include:

- Wetting
- Adsorption and desorption
- Cleaning processes

Wetting characteristics of any solid surface plays an important role in all kinds of different systems. The next most important step is the process of *adsorption* of substances on solid surfaces. These phenomena are the crucial steps for all kinds of *cleaning processes.*

These systems may include the following: washing, coatings, adhesion, lubrication, and oil recovery.

The liquid–solid or $liquid_1$–solid–$liquid_2$ system is both a contact angle (Young's equation) and capillary phenomena (Laplace equation). These two parameters are:

$$\cos(\theta) = (\gamma_S - \gamma_{SL})/\gamma_L$$

and

$$\Delta P = (2\gamma_L \cos(\theta))/\text{Radius} \tag{6.1}$$

In the following sections, we will only consider some important phenomena where these parameters are of importance.

6.2 OIL RECOVERY TECHNOLOGY AND SURFACE FORCES

Current energy demand (approximately 80 million barrels of oil per day, coal, and other forms of sources) is known to have a very high priority with regard to sources, such as oil and gas. Oil is normally found under high temperatures (80°C) and pressures (200 atm) depending on the depth of the reservoir. Hence, oil and gas in reservoirs is produced under high-pressure conditions. The pressure needed depends primarily on the porosity of the reservoir rock and the viscosity of the oil, among

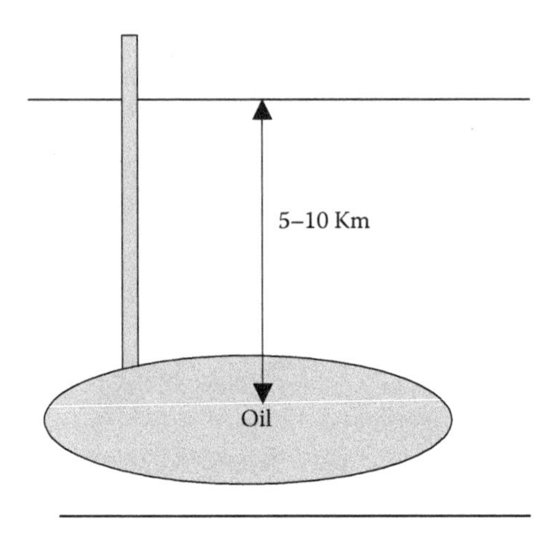

FIGURE 6.1 Oil reservoirs are generally found under large depths, ranging from a few hundred meters to over 10 km (pressure increases at approximately 100 atm/km depth).

other factors (Figure 6.1). This creates the flow of oil through the rocks, which consists of pores of varying sizes and shapes. Roughly, one may compare oil flow to the squeezing of water out of a sponge. The capillary pressure is lower in larger pores than in pores of smaller diameter. Therefore, in the primary oil production one recovers mostly from the larger pores of the reservoir.

The degree of oil recovery from reservoirs is never 100%, and therefore some of the oil in the reservoir remains behind. The major factors arise from capillary forces, as well as adsorption and flow hindrances. The same is valid for oil recovery from shale deposits. This means that all the oil recovered until now leads to some 20% to 40% residual oil in the depleted reservoirs. This may be considered as an advantage in the long run, since as the shortage of oil supplies comes nearer, one may be forced to develop technologies to recover the residual oil (enhanced oil recovery processes, EOR) (Birdi, 2009). The latter subject is being pursued by many research centers around the world.

EOR is a very important research area in the energy supply industry. In all oil reservoirs worldwide a significant amount (over 30%) of oil is not recovered by primary ordinary methods. Studies have shown that by using additives one can recover the residual oil in the reservoirs. Especially those additives that lower the interfacial tension (IFT) give rise to increased oil production. The latter gives increased production when water flooding (or similar method) is used.

In order to enhance this recovery process, physicochemical methods have been used to improve the degree of recovery. In fact, it is safe to conclude that oil recovery processes will need to be developed from these depleted reservoirs, as one approaches the point where the oil becomes scarce. In most oil reservoirs the primary recovery is based on natural flow of oil under the gas pressure of the reservoir. In these reservoirs as this gas pressure drops then water flooding procedure is used. In some reservoirs CO_2 has also been used to increase the oil recovery. The pressure needed is determined by the capillary pressure of the reservoir (which is related to the porosity

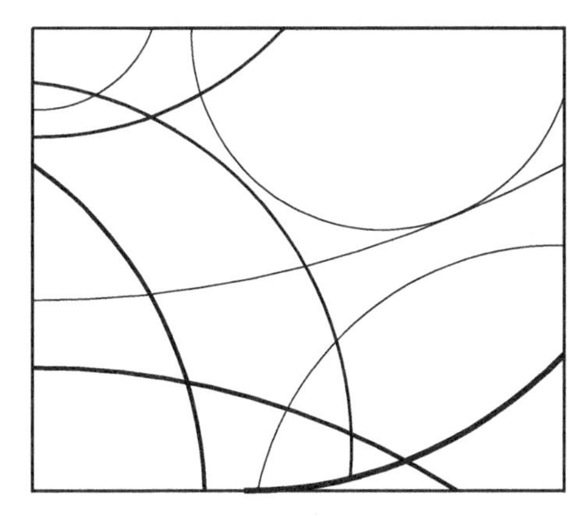

FIGURE 6.2 A porous oil reservoir where capillary forces mainly determine the degree of oil recovery (pores are depicted as dark lines) (see text).

of the reservoir) and the viscosity of the oil. This procedure has been found to leave 30% to 50% of the original oil in the formation. In some cases, substances such as detergents or similar chemicals can be added to enhance the flow of oil through the porous rock structure (Figure 6.2). The principle is to reduce the Laplace pressure (i.e., $\Delta P = \gamma$/curvature) and reduce the contact angle. This process is called *tertiary oil recovery*. The goal is to produce oil that is trapped in a capillary-like structure in the porous oil-bearing material. The addition of surface active agents reduces the oil–water interfacial tension (from about 30 to 50 mN/m to less than 10 mN/m).

The tertiary oil recovery processes are where more complicated chemical additives are designed for a particle reservoir. In all these recovery processes the IFT between oil phase and the water phase is needed. In most reservoirs methods are needed to improve oil recovery. In most operations one uses water (with additives) to push the oil to increase the degree of recovery. However, the matter becomes complex due to various reasons. Another important factor is that during the water flooding the water phase bypasses the oil in the reservoir (Figure 6.3). What this implies

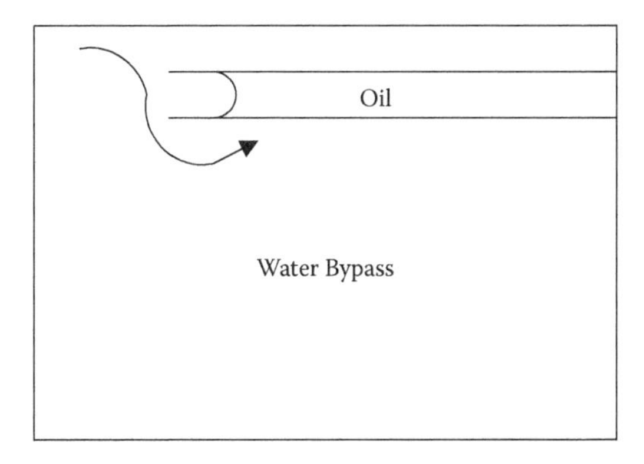

FIGURE 6.3 Water bypass in an oil reservoir. Reservoir pressure is not high enough to push the oil inside the narrow pore.

is that if one injects water into the reservoir to push the oil, most of the water passes around the oil (bypass phenomena) and comes up without being able to push the oil (Birdi, 2002, 2009).

The pressure difference to push the oil drop may be larger than to push the water, thus leading to the so-called bypass phenomena. In other words, as water flooding is performed, due to bypass there is less oil produced while more water is pumped back up with oil. This is a difficult problem to resolve and at this stage requires more research. The use of surfactants and other surface active substances leads to the reduction of $\gamma_{oilwater}$, as described in Figure 6.1. The pressure difference at the oil blob entrapped and the surrounding aqueous phase will be:

$$\Delta P_{oilwater} = 2 \text{ IFT } (1/R_1 \; 1/R_2) \quad (6.2)$$

Thus by decreasing the value of interfacial tension (with the help of surface active agents) (from 50 mN/m to 1 mN/m) the pressure needed for oil recovery would be decreased. In some reservoirs, the addition of CO_2 has also increased the oil production. In the water flooding process mixed emulsifiers are used. Soluble oils are used in various oil-well treating processes, such as the treatment of water injection wells to improve water injectivity, to remove water blockage in producing wells. The same is useful in different cleaning processes on the oil wells. This is known to be effective since water-in-oil microemulsions are found in these mixtures and with high viscosity. The micellar solution is composed essentially of hydrocarbon, aqueous phase, and surfactant sufficient to impart micellar solution characteristics to the emulsion. The hydrocarbon is crude oil or gasoline. Surfactants are alkyl aryl sulfonates, more commonly known as petroleum sulfonates. These emulsions may also contain ketones, esters, or alcohols as cosurfactants. Another capillary phenomena one has to consider is that the pores in the reservoirs are not perfectly circular. In the case of square-shaped pores bypass in the corners have to be considered, which are not found in circular-shaped pores (Birdi et al., 1988).

6.3 APPLICATIONS IN CLEANING PROCESSES

In almost all cleaning processes, unwanted materials (e.g., grease, dirt, color, bacteria) need to be removed from surfaces or cloth (e.g., cotton synthetics, wool). This is a large industry and extensive research based on surface and colloid chemistry can be found in the literature.

6.3.1 Detergency and Surface Chemistry Essential Principles

The detergent industry is a very large and important area where surface and colloid chemistry principles have been extensively applied. In fact, some detergent manufacturers have been involved in highly sophisticated research and development for many decades; some of which are protected by patents.

The procedure for cleaning fabrics or metal surfaces is to primarily remove dirt from the surface of clothes. Second, it is important to ensure that the dirt does not redeposit after its removal. *Dry cleaning* is different since organic solvents are used. The dirt adheres to the fabric through different forces (such as van der Waals and electrostatic). Some components of dirt are water soluble, and some are water insoluble. The detergents used are designed specifically for these particular processes by the industry and the environment. The composition of the soaps or detergents is mainly based upon achieving the following effects:

1. Water should be able to wet the fibers as completely as possible (i.e., the contact angle should be less than 10 degrees). This is achieved by lowering the surface tension, γ, of the washing water, which thus lowers the contact angle. The low value of surface tension also makes the washing liquid to be able to penetrate the pores (if present), since from the Laplace equation the pressure needed would be much lower.

For example, if the pore size of fabric (such as any modern type like microcotton or Gor-Tex) is 0.3 μm, then it will require a certain pressure (= $\Delta P = 2\ \gamma/R$) in order for water to penetrate the fibers. In case of water ($\gamma = 72$ mN/m) and using a contact angle of 105°, we obtain:

$$\Delta P = 2\ (72\ 10^{3}) \cos(105)/0.3\ 10^{6} = 1.4 \text{ bar} \tag{6.3}$$

2. The detergent then interacts with the dirt or soil to start the process of removal from the fibers and dispersion into the washing water. In order to inhibit the soil once removed to readsorb on the clean fiber, polyphosphates or similar suitable inorganic salts are used. These salts also increase the pH (around 10) of the washing water. In some cases, suitable polymeric antiredeposition substances are used (such as carboxymethyl cellulose).
3. After the fabric is clean, special brighteners (fluorescent substances) are used that give a bluish haze to the fabric. This enhances the whiteness (by depressing the yellow tinge). Additionally, these also enhance the color perception. Brighteners used for cotton are different from those used for synthetic fabrics. Hence, the washing process is a series of well-designed steps that the industry has provided with much information as well as state-of-the-art technology. Further, all washing technology has changed all along as the demands have changed. Washing machines are designed to operate in conjunction with the soap industry. The mechanical movement and agitation is coordinated with the soap/detergent characteristics.

Typical compositions of different laundry detergents, shampoo, or dishwashing powder are:

	Laundry	Shampoo	Dishwashing
Detergent (Na-alkyl sulfate)	10–20	25	—
Soaps	5	—	—
Nonionics	5–10	1–5	—
Inorganic salts (polyphosphate, silicates)	3–50	50	—
Optical brighteners	<1	—	—

It is worth noting that the purpose of detergents in these different formulations is different in each case. In other words, detergents today are tailor-made for each specific application (Ruiz, 2008). The detergents in shampoo should provide stable foam in order to increase the cleaning effect, and at the same time there should not be any adverse effect on the structure of hair. On the other hand, laundry detergents for dishwashing should only give a lower surface tension but almost no foaming (because foaming would reduce the cleaning effect). Hence, in dishwashing machine formulations nonionics are used, which are not very soluble in water and thus produce very little (or no) foam. These are sometimes of type EOEOEOPOPOPO (ethylene oxide [EO]–propylene oxide [PO]). The propylene group behaves as *apolar* and the oxide group behaves (through hydrogen bonding) as the *polar* part. These EOPO types can be tailor-made by combining various ratios of EO:PO in the surfactant molecule. In some cases, even butylene-oxide groups have been used. Additionally, in the case of shampoos there are other criteria that have to be strictly controlled. Shampoos are used to wash hair and are designed to remove dirt without damaging the hair (which is mainly composed of proteins and fats). There should also be no skin irritation or eye irritation effects from shampoos. Baby shampoos are particularly manufactured using surfactants that exhibit a minimum amount of eye irritation.

Further, soil consists mainly of particulate, greasy matter, and so forth. The detergents are supposed to keep the soil suspended in the solution and restrict the redeposition. Tests also show that detergents stabilize suspensions of carbon or other solids such as manganese oxide in water. This suggests that detergents adsorb on the particles. Detergents are added redeposition controllers, such as carboxymethylcellulose. The detergents are necessary also to remove the greasy part of soil. The adsorption of detergents on soil particles is involved in the detergency process. In the early age of detergent usage during the 1960s, too much sewage treatment showed foaming problems. Later, detergents were used with better degradation properties and better control. For example, straight-chain alkyls were more biodegradable than branched alkyl chains.

6.3.2 Water Repellency of Materials

In many cases, such as umbrellas and raincoats, material that is nonwetting to water is needed. As mentioned earlier, if the contact angle is larger than 90° then a nonwetting solid surface is present, whereas in the case of less than 90° wetting solid is present.

This is a very general statement but serves as a useful guideline for investigations. However, for water to penetrate fabrics, the magnitude of θ has to be close to zero. This can be achieved by using detergents. However, to achieve water repellency the magnitude of θ has been as large (i.e., >90°) as possible. If $\theta < 90°$, then $\cos(\theta)$ is positive and the liquid would penetrate a fabric. On the other hand if $\theta > 90°$, the sign of $\cos(\theta)$ is negative, and the liquid will not penetrate the material. This should be compared to the capillary rise (or fall) of different liquids in glass tubing (see Chapter 2). In the Young's equation, the quantity $(\gamma_S - \gamma_{SL})$ can be made negative. This is achieved by coating the solid material with some suitable material (such as Teflon). The latter leads to that γ_c should be reduced to less than 30 mN/m.

6.4 EVAPORATION RATES OF LIQUID DROPS

Liquid drops are encountered in many natural (raindrops, fog, river waterfalls) and industrial systems (sprays, oil combustion engines, cleaning processes). The rate of evaporation of liquid from such drops can be of importance in the function of these systems. Extensive investigations on the evaporation of liquid drops (free hanging drops and drops placed on solid surfaces) have been reported in the current literature (Birdi, 2002, 2008, 2010a). These drops have been analyzed as a function of:

- Liquid (water or organic liquids)
- Solids (plastics, glass, etc.)
- Contact angle (θ)
- Height, diameter, and volume
- Weight

In these analyses, some assumptions have been made with regard to the shape of the drops. The most accurate data obtained has been by using the weight method. Different analyses have shown that the rate of evaporation was linearly dependent on the radius of the drop. Further, the contact angle of water drops on Teflon (a nonwetting surface) remained constant under evaporation. On the other hand, the contact angle decreased as water drops evaporated on glass (a wetting surface).

6.5 ADHESION BETWEEN TWO SOLID SURFACES (GLUES)

In industry and many other everyday systems, joining two solids requires using glues or adhesives. For example:

- Plastic on metal (car industry)
- Plastic on glass
- Metal to glass
- Wood to wood (furniture, housing, boats)
- Metal to metal (airplane wings, windmill wings)

The energy needed to break such a contact, say, if we consider the adhesion of plastic on glass, the highest adhesion will be obtained if the adhesive fills all the valleys and crevices of each adhered body surface (Figure 6.4). This will remove any air pockets that do not contribute to adhesion. The role of the adhesive or glue is to provide mechanical interlocking of the adhesive molecules. The strength of the bond is dependent upon the quality of this interlocking interface. For achieving optimum bonding, chemical or physical abrading is used. The abrading process creates many useful properties at the solid surface, including enhancement of the mechanical interlocking, creation of clean surface, formation of a chemically reactive surface, and increase in surface area (a smooth surface has lower surface area than a rough surface).

Diffusion bonding is a form of mechanical interlocking that occurs at the molecular level in polymers.

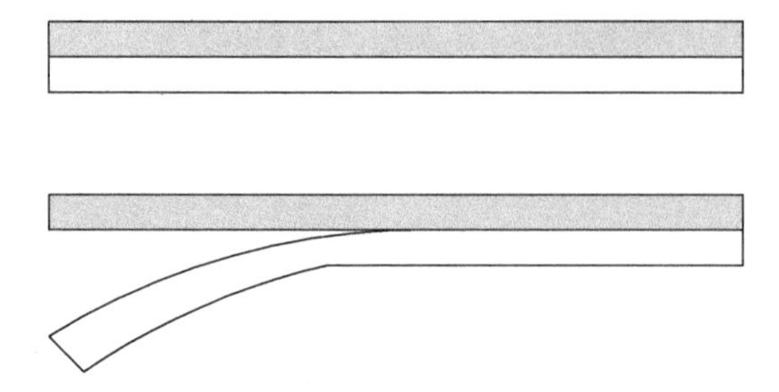

FIGURE 6.4 Adhesion mechanism (peeling process).

The science of *bonding technology* is extensive (Kamperman and Synytska, 2012). A brief description along with some real examples is given in the following. It is important to prime the surfaces of the layers to be bonded, that is, cover the surfaces with a dilute solution of the adhesive mixed with an organic solvent to obtain a dried film thickness between 0.0015 to 0.005 mm and cure separated before applying the adhesive to bond the layers together.

Many theories have been developed to explain the process of bonding in adhesive structures. According to the mechanical bonding theory, an adhesive needs to fill the valleys and crevices of each adherend (body to be bonded) and displace trapped air to work effectively. Adhesion is the mechanical interlocking of the adhesive and the adherend together, and the overall strength of the bond is dependent upon the quality of this interlocking interface. This can be chemical or physical abrading for optimum bonding. Abrading the adherend gives following:

- Enhanced mechanical interlocking
- A clean corrosion-free surface
- A chemically reactive surface
- Increased bonding surface area

Polymers give rise to so-called diffusion bonding, which is a form of mechanical interlocking within large molecules. The adsorption mechanism theory suggests that bonding is the process of intermolecular attraction (van der Waals bonding or permanent dipole, for example) between the adhesive and the adherend at the interface. An important factor in the strength of the bond according to this theory is the wetting of the adherend by the adhesive. Wetting is the process in which a liquid spreads onto a solid surface and is controlled by the surface energy of the liquid–solid interface versus the liquid–vapor and the solid–vapor interfaces. In a practical sense, to wet a solid surface, the adhesive should have a lower surface tension than the adherend. In some systems with *charged surfaces*, the electrostatic forces will have to be considered. Electrostatic forces may also be a factor in the bonding of an adhesive to an adherent. These forces arise from the creation of an electrical double layer of separated charges at the interface and are believed to be a factor in the resistance to separation of the adhesive and the adherend. An adhesive is the basic substance that, when brought into contact with the hardener, after a chemical reaction, the former becomes an effective adherent.

This theory has been developed to explain the curious behavior of the failure of bonded materials. Upon failure, many adhesive bonds break not at the adhesion interface, but slightly within the adherend or the adhesive, adjacent to the interface. This suggests that a boundary layer of weak material is formed around the interface between the two media. In the following some mechanisms of adhesive failure are described. It has been found that two predominant mechanisms of failure exist in adhesively bonded joints:

1. Adhesive failure
2. Cohesive failure

Adhesive failure is the interfacial failure between the adhesive and one of the adherends. It indicates a weak boundary layer often from improper surface preparation or adhesive choice. Cohesive failure is the internal failure of either the adhesive or, rarely, one of the adherends.

Ideally, the bond will fail within one of the adherends or the adhesive. This indicates that the maximum strength of the bonded materials is less than the strength of the adhesive strength between them. Usually, the failure of joints is neither completely cohesive nor completely adhesive. It is thus obvious, that for good bonding the surfaces need to be clean, and any dirt, grease, and lubricants, water or moisture, and weak surface scales need to be removed. Solvents are used to clean the soil from solid surfaces from the adherend using an organic solvent without affecting any physical property.

The following cleaning procedures have been found to be useful:

- Vapor degreasing
- Solvent wiping or immersion or spraying
- Ultrasonic vapor degreasing

Ultrasonic treatment with a subsequent solvent rinse of the surface is the most convenient method. Other intermediate procedures, such as abrasive scrubbing, filing, or detergent cleaning can also be used. Cleaning can be done with a chemical treatment, which includes acid or alkaline etching of the adherend surface. Especially, the etching process removes stubborn oxides (such as on Al or Fe) and roughens the surface on a microscopic scale. Priming protects the surface from oxidation, improves wetting, helps prevent adhesive peeling, and serves as a barrier layer to prevent undesirable reactions between the adhesive and the adherent. An adhesive is the basic substance that, when brought into contact with the hardener after a chemical reaction, becomes an effective adherent. These are solvents, which reduce the thickness of adhesive and also penetrate the surfaces. In some cases, fillers are used to enhance adhesion and reduce costs.

7 Colloidal Dispersion Systems: Physicochemical Essential Properties

7.1 INTRODUCTION

Solid particles dispersed in water are found in many examples in daily life, such as wastewater treatment. In this, chapter, the essential aspects in the very large industrial application of *colloid chemistry* will be described. Surprisingly, mankind has been aware of colloids for many thousands of years. Civilizations such as ancient Egypt and Maya civilizations used their knowledge about adhesion (between blocks of stones) when building pyramids thousands of years ago. This was long before modern-day cement was invented. Even the mud houses man has built were based upon the behavior of colloid aspects of materials used in such processes, such as clay and cow dung. In everyday life, one comes across solid particles of different sizes, ranging from stones on a beach, sand particles, or dust floating in the air. A special relation between particle size (surface area) and their characteristics exists. The rather small particles which range in size from 50 Å to 50 μm are called *colloids*. The simplest difference is when sand particles versus dust particles are considered. It is almost fascinating to observe how dust or other fine particles remain in suspension in air. On occasion, it has been observed that a particle gets a collision-like thrust. Already in the 19th century (Brown), it was observed under the microscope that small microscopic particles suspended in water made some erratic movements (as if hit by some other neighboring molecules) (Adamson and Gast, 1997). This has since then been called *Brownian motion*. The erratic motion arises from the kinetic movement of the surrounding water molecules. Thus, colloidal particles remain suspended in solutions through Brownian motion, only if gravity forces do not drag these to the bottom (or top). If one throws some sand into air, the particles fall to the earth rather quickly. On the other hand, in the case of talcum particles, these continue to float in the air for a long time.

These characteristics will be described here. The size of particles may be considered from the following data:

Colloidal Dispersions	Size Range (nm–10 μ)
Mist/fog	0.1 μ–10 μ
Pollen/bacteria	0.1 μ–10 μ
Oil in smoke/exhaust	1 μ–100 μ
Virus	1 nm–10 μ
Polymers/macromolecules	0.1 nm–100 nm
Micelles	0.1 nm–10 nm
Vesicles	1 μ–1000 μ

This shows that there is a range in size of particles. Actually, in colloids it is the size of particles that is of primary interest. The stability of such colloidal systems is something that may be compared to whether the system stays energetically stable, or it will take up a new state of more stable configuration. This may be roughly compared to a bucket that is stable when standing up, but if tilted beyond a certain angle, it topples and comes to rest on its side (Figure 7.1).

A colloidal suspension may be *unstable* and exhibit separation of particles within a very short time. Or it may be *stable* for a very long time, such as over a year or more. And there will thus be found a *metastable* state, which would be in between these two. This is an oversimplified example, but it shows that one should proceed to analyze any colloidal system following these three criteria. The most remarkable finding that can be mentioned about colloidal suspensions is that these systems can exist at all! Especially, since some solid suspensions can be stable for a very long time. In pharmaceutical applications, one important example is the use of suspension of insulin in pen injections. The insulin suspension is stable for long enough time for its application, which provides very accurate dosage to the patient. A wide variety of pharmaceutical products that are based on suspended molecules exists.

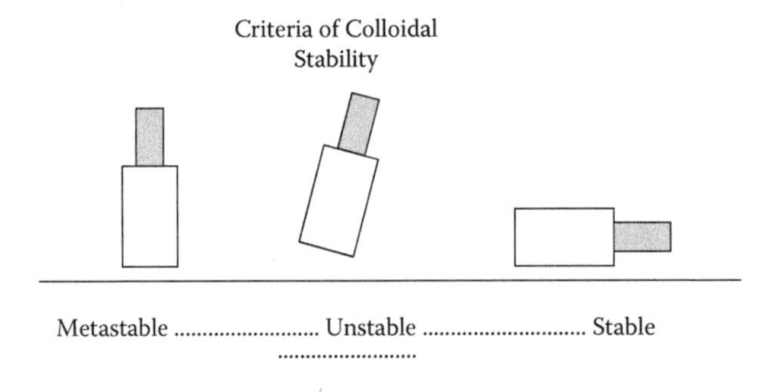

FIGURE 7.1 Stability criteria of any colloidal system: metastable, unstable, and stable states.

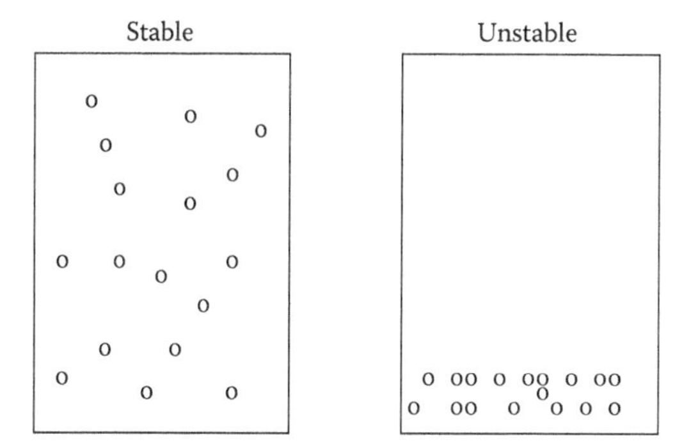

FIGURE 7.2 Stable wastewater suspension of particles and unstable suspension after suitable pH adjustment and other additives. (See text for details.)

As an example, consider the wastewater treatment process. Wastewater with colloidal particles is a stable suspension. However, by treating it with some definite methods (such as pH control or electrolyte concentration) the stability of the system can be changed, as shown in Figure 7.2.

Wastewater treatment technology is one of the most important areas of surface chemistry applications. The suspended materials are separated by coagulation and filtered away. The soluble pollutants are removed by other procedures.

In colloidal systems, van der Waals forces play an important role. When any two particles (neutral or with charges) come very close to each other, the van der Waals forces will be strongly dependent on the surrounding medium; in a vacuum two identical particles will always exhibit attractive force. On the other hand, if two different particles are present in a medium (in water), then there may be repulsion forces. This can be due to one particle adsorbing with the medium more strongly than with the other particle. One example will be silica particles in water medium and plastics (as in wastewater treatment). It is important to understand under what conditions it is possible that colloidal particles remain suspended. Especially, if paint aggregates in the container, then it is obviously useless.

When solid (inorganic) particles are dispersed in aqueous medium, ions are released in the medium. The ions released from the surface of the solid are of opposite charge.

This can be easily shown when glass powder is mixed in water—conductivity will increase with time (this can be easily measured in the case of powdered glass). The presence of same charge on particles in close proximity gives repulsion, which keeps the particles apart (Figure 7.3).

The charged positive–positive (or negative–negative) particles will show repulsion. On the other hand, the positive–negative particles will attract each other. The ions distribution will also depend on the concentration of any counterions or co-ions in the solution. Even glass when dipped in water exchanges ions with its

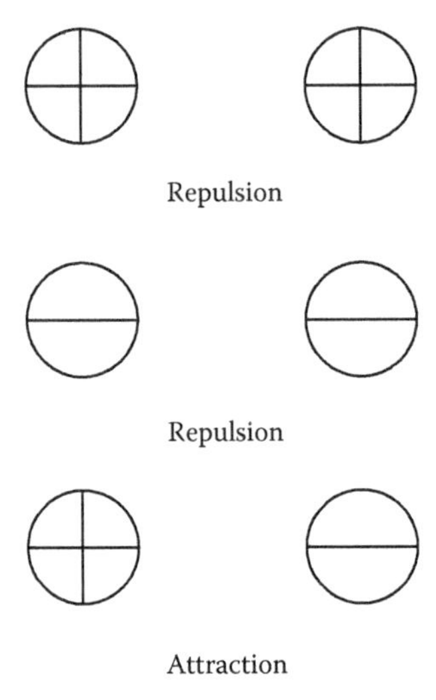

FIGURE 7.3 Solid particles with charges: positive–positive (repulsion), negative–negative (repulsion), and positive–negative (attraction).

surroundings. Such phenomena can be easily investigated by measuring the change in conductivity of the water.

The force, F_{12}, acting between opposite charges is given by Coulomb's law, with charges q_1 and q_2, separated at a distance R_{12}, in a dielectric medium, D_e:

$$F_{12} = (q_1\ q_2)/(4\ \pi\ D_e \varepsilon_o R_{12}) \tag{7.1}$$

where the force would be attractive between opposite charges, whereas repulsive in the case of similar charges. Since D_e of water is very high (80 units) as compared to D_e of air (about 2), we will expect very high dissociation in water, while hardly any dissociation in air or organic liquids. Let us consider the F_{12} for Na^+ and Cl^- ions (with charge of $1.6\ 10^{-19}$C = $4.8\ 10^{-10}$ esu) in water (D_e = 74.2 at 37°C), and at a separation (R_{12}) of 1 nm:

$$F_{12} = [(1.6\ 10^{-19})(1.6\ 10^{-19})]/[(4\ \Pi\ 8.854\ 10^{-12})\ (10^{-9})(74.2)] \tag{7.2}$$

where ε_o is $8.854\ 10^{-12}\ kg^{-1}m^{-3}\ s^4\ A^2$ ($J^{-1}\ C^2\ m^{-1}$). This gives a value of F_{12} of $^-3.1$ 10^{-21} J/molecule or $^-1.87$ kJ/mole.

Another important physical parameter that should be considered is the *size distribution* of the colloids. A system consisting of particles of the same size is called a monodisperse. A system with different sizes is called polydisperse. It is also obvious that monodisperse systems will exhibit different properties than those polydisperse. In many industrial applications (such as coating of substances in colloid form on tapes, as used for recording music, and coatings on CDs or DVDs), the size of particles is an important characteristic. The methods used to prepare monodisperse colloids is to achieve a large number of critical nuclei in a short interval of time. This induces all equally sized nuclei to grow simultaneously and thus producing a monodisperse colloidal product.

7.2 COLLOID STABILITY (DLVO THEORY)

It is important to understand under which conditions a colloidal system will remain dispersed (and under which other conditions become unstable). How colloidal particles interact with each other is one of the important problems that determines the understanding of the experimental results for phase transitions in such system as found in various industrial processes. One also will need to know under which conditions a given dispersion will become unstable (*coagulation*). For example, coagulation needs to be applied in wastewater treatment so that most of the solid particles in suspension can be removed. When any two particles come close to each other, different forces exist (depending on the distance between the particles): *attractive forces* and *repulsive forces.*

If the attractive forces are larger than the repulsive forces, then the two particles will merge together. However, if the repulsion forces are larger than the attractive forces, then the particles will remain separated. It is important to mention here that the medium in which these particles are present thus will to some degree contribute also. Especially, it has been found that pH and ionic strength (i.e., concentration of ions) exhibit very specific effects on colloidal stability.

The different forces of interest are:

- Van der Waals
- Electrostatic
- Steric
- Hydration
- Polymer–polymer interactions (if polymers are involved in the system)

In many systems, large molecules (polymers) can be added, which when adsorbed on the solid particles will impart a special kind of stability criteria. It is well known that neutral molecules, such as alkanes, attract each other mainly through van der Waals forces. Van der Waals forces arise from the rapidly fluctuating dipoles moment (10^{15} sec^{-1}) of a neutral atom, which leads to polarization and consequently to attraction. This is also called the London potential between two atoms in a vacuum and is given as:

$$V_{vdw} = -(L_{11}/R^6) \tag{7.3}$$

where L_{11} is a constant that depends on the polarizability and the energy related to the dispersion frequency, and R is the distance between the two atoms. Since the London interactions with other atoms may be neglected as an approximation, the total interaction for any macroscopic bodies may be estimated by a simple integration.

When two similarly charged colloid particles, under the influence of the electrical double layer (EDL), come close to each another they will begin to interact. The potentials will feel each other and this will lead to consequences. The charged molecules or particles will be under both vw and electrostatic interaction forces. The van der Waals forces that operate at short distance between particles will give rise to strong attraction forces. The potential of mean force between colloid particles in electrolyte solutions plays a central role in the description of the phase behavior and the kinetics of agglomeration in colloidal dispersions. This kind of investigation is important in various industries:

- Inorganic materials (ceramics and cements)
- Foods (milk)
- Biomacromolecular systems (proteins and DNA)

DLVO (Derjaguin–Landau–Verwey–Overbeek) theory describes that the stability of a colloidal suspension is mainly dependent on the distance between the particles (Adamson and Gast, 1997; Grodzka and Pomianowski, 2005; Birdi, 2009, 2010a,b). DLVO theory has been modified in later years and different versions are found in the current literature.

The electrostatic forces give rise to repulsion at large distances (Figure 7.4). This arises from the fact that the electrical charge–charge interactions take place at a large distance of separation. The resultant curve is shown in Figure 7.3. The barrier height

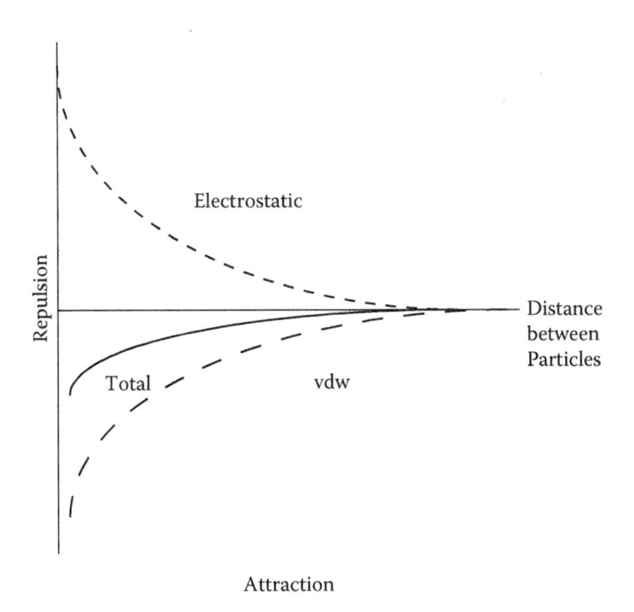

FIGURE 7.4 Variation of repulsion and attraction forces versus distance between two particles.

determines the stability with respect to the quantity k T, the kinetic energy. DLVO theory predicts, in the simplest terms, that if the repulsion potential (Figure 7.4) exceeds the attraction potential by a value

$$W >> k\,T \tag{7.4}$$

then the suspension will be stable. On the other hand, if

$$W <= k\,T \tag{7.5}$$

then the suspension will be unstable and it will coagulate. It must be stressed that DLVO theory does not provide a comprehensive analyses. It is basically a very useful tool for such analyses of complicated systems. It is an especially useful guidance theory in any new application or any industrial development.

7.2.1 Charged Colloids (Electrical Charge Distribution at Interfaces)

In everyday life, electrically charged particles or surfaces play a very important role. The interactions between two charged bodies will be dependent on various parameters (e.g., surface charge, electrolyte in the medium, charge distribution) (Figure 7.5). The distribution of ions in an aqueous medium needs to be investigated in such charged colloidal systems. This observation means that the presence of charges on surfaces indicates that a potential exists that needs to be investigated. On the other hand, in the case of neutral surfaces one has only the van der Waals forces to consider. This was clearly seen in the case of micelles, where the addition of small amounts of NaCl to the solution showed:

- A large decrease in critical micelle concentration (CMC) in the case of ionic surfactant
- Almost no effect in nonionic micelles (since in these micelles there are no charges or EDL)

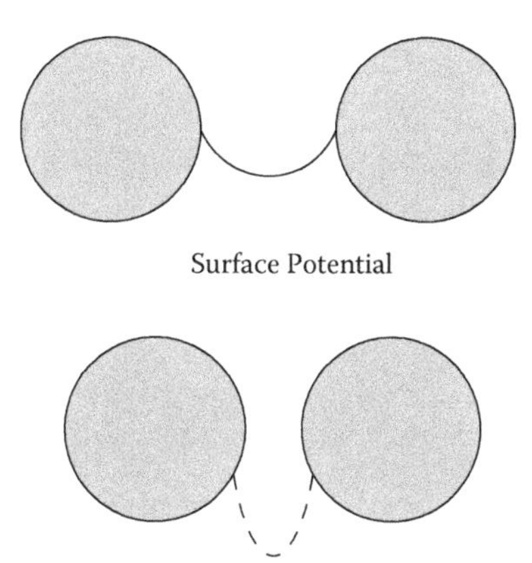

FIGURE 7.5 Variation of EDL between two charged particles with different ion concentrations (low, solid line; high, dashed line).

The addition of electrolytes produces a different kind of surface potential curve. This is easily verified in applications such as washing clothes.

Electrostatic and EDL forces play an important role in a variety of systems known in science and engineering (Birdi, 2010b). It is useful to consider a specific example to understand these phenomena. Let us take a surface with positive charge, which is suspended in a solution containing positive and negative ions. There will be a definite surface potential, ψ_o, which decreases to a value zero as one moves away into solution (Figure 7.6).

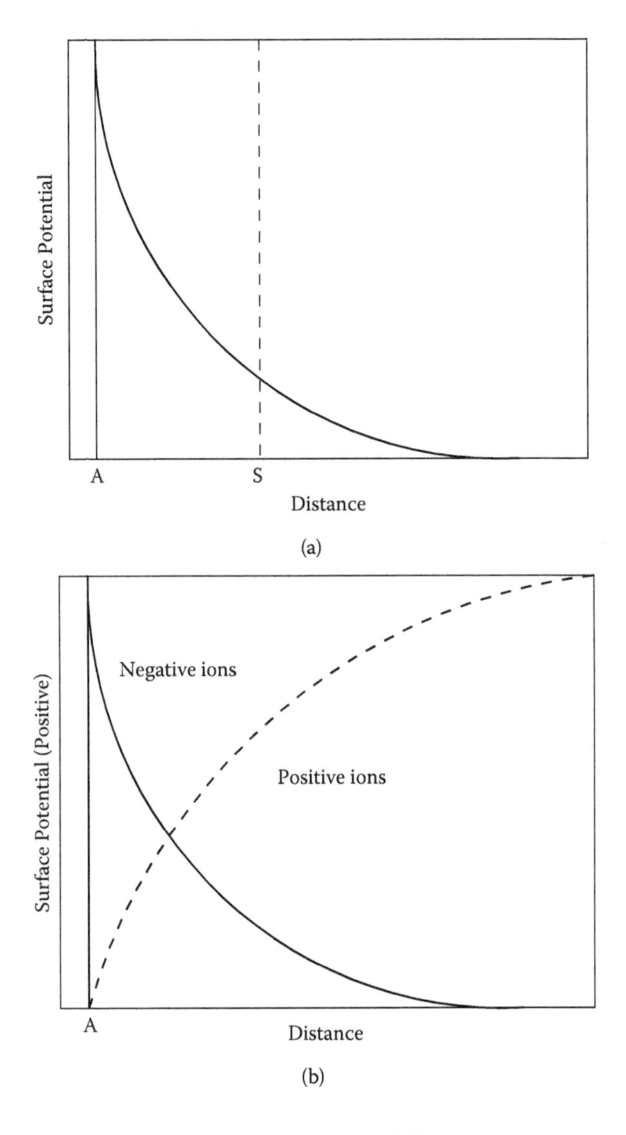

FIGURE 7.6 Variation of (a) surface potential, ψ_o, and (b) ions (sodium and chloride) versus distance from the solid with a positive surface charge.

It is obvious that the concentration of positive ions will decrease as one approaches the surface of the positively charged surface (charge–charge repulsion). On the other hand, the oppositely charged ions, negative, will be strongly attracted toward the surface. This gives rise to the so-called Boltzmann distribution:

$$n^- = n_o\, e^{(z\, \varepsilon\, \psi/k\, T)} \tag{7.6}$$

$$n^+ = n_o\, e^{-(z\, \varepsilon\, \psi/k\, T)} \tag{7.7}$$

This shows that positive ions are repelled, while negative ions are attracted to the positively charged surface. At a reasonable far distance from the particle, $n^+ = n^-$ (as required by the electroneutrality). Through some simple assumptions, one can obtain an expression for ψ (*r*), as a function of distance, *r*, from the surface as:

$$\psi\, (r) = z\, e/(D\, r)\, \varepsilon - \kappa\, r \tag{7.8}$$

where κ is related to the ion atmosphere around any ion. In any aqueous solution when an electrolyte, such as NaCl, is present, it dissociates into positive (Na^+) and negative (Cl^-) ions. Due to the requirement of electroneutrality (that is there must be same positive and negative ions), each ion is surrounded by an appositively charged ion at some distance. This distance will decrease with increasing concentration of the added electrolyte. The expression $1/\kappa$ is called the Debye length. As expected, the Debye-Huckel theory tells us that ions tend to cluster around the central ion. A fundamental property of the counterion distribution is the thickness of the ion atmosphere (Birdi, 2010b). This thickness is determined by the quantity Debye length or Debye radius ($1/\kappa$). The magnitude of $1/\kappa$ has dimension in centimeters, as follows:

$$\kappa = [(\varepsilon_r\, \varepsilon_o\, k_B T)/(2\, N_A\, e^2\, I)]^{1/2} \tag{7.9}$$

The values of $k_B = 1.38\ 1^{-23}$ J/molecule K, e = 4.8 10^{-10} esu, ε_r is the dielectric constant, and ε_o is the permittivity of free space. Thus the quantity $k_B T/e = 25.7$ mV at 25°C. As an example, an 1:1 ion (such as NaCl or KBr) with concentration 0.001M, one gets the value of $1/\kappa$ at 25°C (298 K):

$$\begin{aligned} 1/\kappa &= (78.3\ 1.38\ 10^{-16}\ 298)/(2\ 4\ \Pi\ 6.023\ 10^7) \\ &\quad [(4.8\ 10^{-10})^2]^{0.5} \\ &= 9.7\ 10^{-7}\ \text{cm} \\ &= 97\ \text{Å} \end{aligned} \tag{7.10}$$

The expression in Equation (7.8) can be rewritten as:

$$\psi\, (r) = \psi_o\, (r)\, \exp(-\, \kappa\, r) \tag{7.11}$$

which shows the change in ψ (r) with the distance between particles (r). At a distance $1/\kappa$ the potential has dropped to ψ_o. This is accepted and corresponds with the thickness of the double layer. This is the important analysis, since the particle–particle

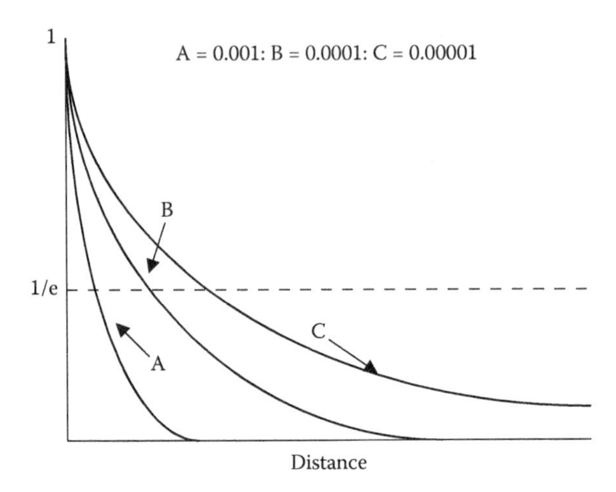

FIGURE 7.7 Variation (decrease) in electrostatic potential with distance of separation as a function of electrolyte concentration (ionic strength).

interaction is dependent on the change in ψ (r). The decrease in ψ (r) at the Debye length is different for different ionic strength (Figure 7.7 and Table 7.1).

The data in Table 7.1 shows the values of the Debye-Huckel radii in various salt concentrations. The magnitude of $1/\kappa$ decreases with λ and with the number of charges on the added salt. This means that the thickness of the ion atmosphere around a reference ion will be much compressed with increasing value of λ and the magnitude of z_{ion}.

A trivalent ion such as Al^{3+} will compress the double layer to a greater extent in comparison with a monovalent ion such as Na+. Further, inorganic ions can interact with charge surface in one of two distinct ways:

1. Nonspecific ion adsorption where these ions have no effect on the isoelectric point
2. Specific ion adsorption, which gives rise to change in the value of the isoelectric point

---

TABLE 7.1
Magnitude of the Debye Length ((1/*k*) nm) in Aqueous Solutions

Salt Concentration (molal)	1:1	1:2	2:2
0.0001	30.4	17.6	15.2
0.001	9.6	5.55	4.81
0.01	3.04	1.76	1.52
0.1	0.96	0.55	0.48

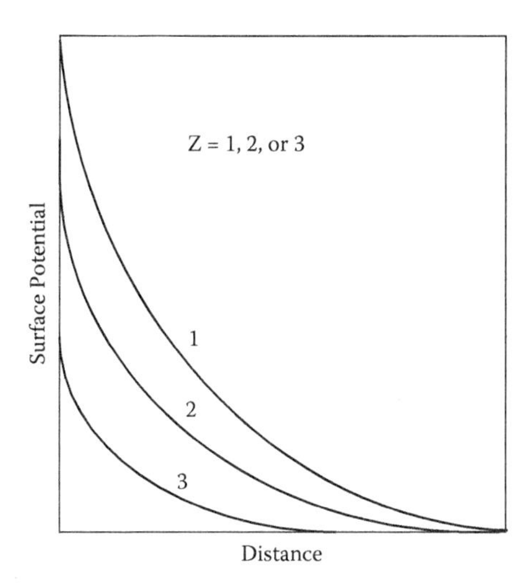

FIGURE 7.8 Variation of the diffuse double layer as a function of charge on the ions (Z).

Under those conditions where the magnitude of $1/\kappa$ is very small (for example, in high electrolyte solution), one can write:

$$\psi = \psi_o \exp -(\kappa x) \tag{7.12}$$

where x is the distance from the charged colloid.

The value of ψ_o is 100 mV (in the case of monovalent ions) (= $4\, k_B\, T/z\, e$).

Experimental data and theory show that the variation is dependent on the concentration and the charge of the ions (Figure 7.8).

These data show:

- The surface potential drops to zero at a faster rate if the ion concentration (C) increases.
- The surface potential drops faster if the value of z goes from 1 to 2 or larger.

In washing powders, multicharged phosphates for example, are used to enable dirt particles to stay off the clean fabrics.

7.2.2 Colloidal Electrokinetic Processes

Charged colloids allow the visual investigation of these systems under dynamic conditions. In the following, let us consider what happens if the charged particle or surface is under dynamic motion of some kind. Further, there are different systems under which the electrokinetic phenomena are investigated. These systems are:

1. *Electrophoresis*—This system refers to the movement of the colloidal particle under an applied electric field. In biology, different proteins exhibit different charges, and thus can be separated using this property.

- Negatively charged particle moves toward the positive electrode.
- Positively charged particle moves toward the negative electrode.

- The speed of movement of a charged particle is dependent on various parameters:
  - Number of charges
  - Size and shape of particle

Thus, it is seen that particles can be separated by using the electrophoretic technique.

---

2. *Electroosmosis*—This system is where a fluid passes next to a charged material. This is actually the complement of electrophoresis. The pressure needed to make the fluid flow is called the *electroosmotic pressure.* Fluid movement through a charged material (such as earth) gives rise to electroosmotic pressure. This arises from asymmetrical charge distribution at the liquid–solid interface, which depends on the magnitude of the surface potential.
3. *Streaming potential*—If fluid is made to flow past a charged surface then an electric field is created, which is called streaming potential. This system is thus opposite of electroosmosis.
4. *Sedimentation potential*—A potential is created when charged particles settle out of a suspension. This gives rise to sedimentation potential, which is the opposite of the streaming potential. The reason for investigating electrokinetic properties of a system is to determine the quantity known as the zeta potential.

Electrophoresis is the movement of an electrically charged substance under the influence of an electric field. This movement may be related to fundamental electrical properties of the body under study and the ambient electrical conditions in Equation (7.13). F is the force, q is the charge carried by the body, E is the electric field:

$$F_e = q\,E \tag{7.13}$$

The resulting electrophoretic migration is countered by forces of friction such that the rate of migration is constant in a constant and homogeneous electric field:

$$Ff = v\,f_r \tag{7.14}$$

where v is the velocity and f_r is the frictional coefficient.

$$Q\,E = v\,f_r \tag{7.15}$$

The electrophoretic mobility μ is defined as followed.

$$\mu = v/E = q/f_r \tag{7.16}$$

The preceding expression is only applied to ions at a concentration approaching 0 and in a nonconductive solvent. Polyionic molecules are surrounded by a cloud of counterions that alter the effective electric field applied on the ions to be separated. This renders the previous expression, which is a poor approximation of what really happens in an electrophoretic apparatus.

The mobility depends on both the particle properties (e.g., surface charge density and size) and solution properties (e.g., ionic strength, electric permittivity, and pH).

For high ionic strengths, an approximate expression for the *electrophoretic mobility*, μ_e, is given by the Smoluchowski equation:

$$\mu_e = \varepsilon\, \varepsilon_o\, \eta/\zeta \tag{7.17}$$

where ε is the dielectric constant of the liquid, ε_o is the permittivity of free space, η is the viscosity of the liquid, and ζ is the zeta potential (i.e., surface potential) of the particle.

7.2.3 Stability Criteria of Lyophobic Suspensions: Critical Flocculation Concentration and the Schultze-Hardy Rule

Depending on various parameters, solid particles will exhibit a varying degree of stability. Thus, it is interesting to determine how these systems are stabilized with regard to the various forces interacting between the particles.

Solids in suspension can separate out of solution in various stages and pathways. The two most common pathways are:

1. Stable suspension → Flocculation → Coagulation → Sedimentation → Particle separation
2. Stable suspension → Partial sedimentation → Flocculation → Coagulation → Particle separation

The most important system for mankind is wastewater treatment. However, natural phenomena occurring around rivers, lakes, and oceans are also of prime interest in ecology and future life on earth.

Particles in all kinds of suspensions or dispersions interact with two different kinds of forces: attractive forces and repulsive forces. Lyophobic suspensions (sols) must exhibit a maximum in repulsion energy to have a stable system. The total interaction energy, V(h), is given as (Chattoraj and Birdi, 1984; Gisler et al., 1994; Adamson and Gast, 1997; Birdi, 2002, 2009):

$$V(h) = V_{el} + V_{vdw} \tag{7.18}$$

where V_{el} and V_{vdw} are electrostatic repulsion and van der Waals attraction components. Dependence of the interaction energy V(h) on the distance, h, between particles has been ascribed to coagulation rates as follows:

- During slow coagulation
- When fast coagulation sets in

The dependence of energy on h and V(h),

$$V(h) = (64\ C\ RT\ \psi^2)/k \exp(-k\ h) - H/(2\ h^2) \tag{7.19}$$

satisfies the requirements of this coagulation rate. For a certain ratio of constants it has the shape shown in Figure 7.9. For large values of h, V(h) is negative (attraction), following the energy of attraction V_{vdw}, which decreases more slowly with increasing distance (~1/h²). At short distances (small h), the positive component V_{el} (repulsion), which increases exponentially with decreasing h, (exp(−k h), can overcompensate V_{vdw} and reverse the sign of both dV(h)/dh and V(h) in the direction of repulsion. On further reduction of the gap (very small h), V_{vdw}, should again predominate, since:

$$V_{el} = 64\ C\ RT\ \psi^2/k, \text{ as } h \to 0 \tag{7.20}$$

whereas the magnitude of V_{vdw} increases indefinitely when h > 0. There is thus a repulsion maximum in the function V(h), which can be easily found from the condition dV(h)/dh = 0. Figure 7.9 shows the transition from stability a to instability c as the electrolyte concentration, C, is increased. Curve b corresponds to the onset of rapid coagulation. The choice of solution (maximum or minimum) does not present any difficulty since V(h) is positive for the maximum. The solution of Equation (7.20) with respect to V_{max} is involved and will not be discussed here. It is, however, readily appreciated that when the electrolyte concentration is increased, the magnitude of k in the exponent of V_{el} also increases (compression of diffuse layers), so that the maximum caused by it becomes lower. At a certain value of C, the curve V(h) will become similar to curve b in Figure 7.9 with $V_{max} = 0$.

In accordance with all that has been said before, coagulation will become fast starting from this concentration. This is therefore the *critical concentration*, C_{cc}. In other words, the critical concentration can be estimated from simultaneous solution of the following:

$$dV(h)/dh = 0 \quad \text{and} \quad V(h) = 0 \tag{7.21}$$

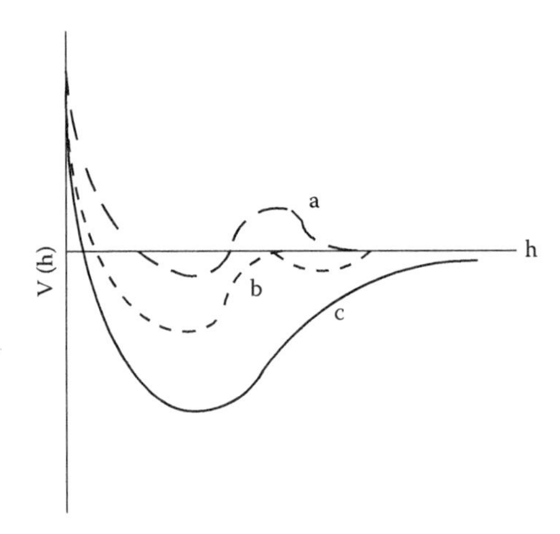

FIGURE 7.9 Variation of electrostatic potential, V(h), is a function of distance.

One can write the following:

$$dV(h)/dh = [-(64\ C_{cc}\ RT\ \psi^2)/k\ \exp(-k_{cr}\ h_{cr}) + K/(h_{cr}^{\ 3})] = 0 \qquad (7.22)$$

and

$$V(h) = [(64\ C_{cc}\ RT\ \psi 2)/k_{cr}\ exp(-k_{cr}\ h_{cr})] - K/(2\ h_{cr}\ 2)] = 0 \qquad (7.23)$$

After expanding these expressions, as related to h and C, this becomes (Schulze-Hardy rule for solid suspensions in water),

$$C_{cc} = 8.7\ 10^{-39}/Z^6\ A^2$$

$$C_{cc}\ Z^6 = \text{constant} \qquad (7.24)$$

where the constant includes (Hamaker constant is approximately 4.2 10^{-19} J) all quantities except Z. This shows that critical concentrations of ions to the sixth power of various valencies are inversely proportional to valency

$$Z = 1 : (2^6)\ 0.016 : (3^6)\ 0.0014 : (4^6)\ 0.000244 \qquad (7.25)$$

The flocculation concentrations of mono-, di- and trivalent gegenions should from this theory expected as:

$1 : (1/2)^6 : (1/3)^6$
! : 1/64 : 1/729

It then becomes obvious that the colloidal stability of charged particles is dependent on:

- Concentration of electrolyte
- Charge on the ions
- Size and shape of colloids
- Viscosity

The critical concentration (critical coagulation concentration) is thus found to depend on the type of electrolyte used as well as on the valency of the counterion. It is seen that divalent ions are 60 times as effective as monovalent ions. Trivalent ions are several hundred times more effective than monovalent ions. However, ions that specifically adsorb (such as surfactants) will exhibit different behavior. Based on these observations, in washing powders composition, one has used multivalent phosphates, for instance, to keep the charged dirt particles from attaching to the fabrics after having been removed. Another example is the wastewater treatment, where for coagulation purposes one uses multivalent ions.

The interface of a mineral (rock) in contact with an aqueous phase exhibits a surface charge. The currently accepted model of this interface is the EDL model of Stern. Chemical reactions take place between the minerals and the electrolytes in the aqueous phase, which results in net charge on the mineral. Water and electrolytes bound to the rock surface constitute the Stern (or Helmholtz) layer. In this region the ions are tightly bound to the mineral, while away from this layer (the so-called diffuse layer), the ions are free to move about. Since the distribution of ions (positive and negative) is even in the diffuse region, there is no net charge. On the other hand, in the Stern layer there will be asymmetric charge distribution, and thus one will measure from zeta-potential data that the mineral exhibits a net charge.

7.3 KINETICS OF THE COAGULATION OF COLLOIDS

Colloidal solutions are characterized by the degree of stability or instability. This is related to the fact that it is important to understand both kinds of properties in everyday phenomena. The kinetics of coagulation is studied by using different methods. The number of particles, N_p, at a given time is dependent on the diffusion-controlled process. The rate is given by:

$$-d\,N_p/dt = 8\pi DRN_p^2 \quad (7.26)$$

where D is the diffusion coefficient and R is the radius of the particle. The rate can rewritten as:

$$-d\,N_p/dt = 4\,k_BT/3\,v\,N_p^2 = k_o N_p^2 \quad (7.27)$$

where $D = k_BT/6\pi vR$, the Einstein equation is applied, and k_o is the diffusion-controlled constant. In real systems one is both interested in stable colloidal systems (as in paints and creams), whereas in other cases one is interested in unstable systems (as in wastewater treatment).

Thus, it is seen from DLVO considerations that the degree of colloidal stability will be dependent on the following factors:

- Size of particles; larger particles will be less stable
- Magnitude of surface potential
- Hamaker constant (H)
- Ionic strength
- Temperature

The attraction force between two particles is proportional to the distance of separation and a Hamaker constant (specific to the system). The magnitude of the Hamaker constant is of the order of 10^{-12} erg (Adamson and Gast, 1997; Birdi, 2002).

Thus, DLVO theory is useful for predicting and estimating colloidal stability behavior. In such systems with many variables, this simplified theory is to be expected to fit all kinds of systems. In the past decade, much development has taken place with regard to measuring the forces involved in these colloidal systems. In one method, the procedure was used to measure the force present between two solid surfaces at very low distances (less than micrometer). The system can operate underwater and thus the effect of additives has been investigated. These data have provided verification of many aspects of DLVO theory. Recently, atomic force microscopy (AFM) has been used to directly measure these colloidal forces (Birdi, 2002). Two particles are brought closer and force (nano-Newton) is measured. Commercially available apparatus are designed to perform such analysis. The measurements can be carried out in fluids and under various experimental conditions (such as added electrolytes and pH).

7.3.1 Flocculation and Coagulation of Colloidal Suspensions

It is known from common experience that a colloidal dispersion with smaller particles is more stable than one with larger particles. The phenomenon of smaller particles forming aggregates with larger size is called *flocculation* or *coagulation.* For example, to remove insoluble and colloidal metal precipitated one uses flocculation. This is generally achieved by reducing the surface charges, which gives rises to weaker charge–charge repulsion forces. As soon as the attraction forces (vdw) become larger than the electrostatic forces then coagulation takes place. Coagulation is initiated by particle charge neutralization (by changing pH or other methods, such as charged poly-electrolytes), which leads to aggregation of particles to form larger size. This approach is based on changing from a charged particle to a neutral particle.

This means that:

- Initial state—Charge–charge repulsion
- Final state—Neutral–neutral (attraction)

Coagulation can be also brought by adding suitable substances (coagulants) particular for a given system. They reduce the effective radius of the colloid particle and lead to coagulation. Flocculation is a secondary process after coagulation, and this leads to very large particle (floccs) formation. Experiments show that coagulation takes place when the zeta potential is around ±0.5 mV. Coagulants such as iron and aluminum inorganic salts are effective in most cases. In wastewater treatment plants, the zeta potential is used to determine the coagulation and flocculation phenomena. The magnitude of zeta potential can be varied by changing the pH. Most of the solid material in wastewater is negatively charged.

7.4 DISPERSION OF SOLID PARTICLES IN FLUIDS

As described earlier, when a solid particle or a liquid drop is broken down in size, then the free energy of the system increases (because the magnitude of surface area per gram of solid increases). This arises from the fact that molecules in the bulk phase are brought to the surface, which needs energy (work). The change in surface free energy is the product of the surface area produced (surface area increase) and the surface tension of the interface created. It also means that by changing the surface tension, different sized particles for a given amount of energy input (such as grinding or rolling or shaking) may be produced. In the grinding process, mechanical energy breaks down the particles. If surface tension is decreased (by adding a suitable surface active agent) then for the same input of mechanical energy the size of particles will decrease with the decrease in surface tension. Moreover, one also finds in practice that if the grinding is performed under dry conditions, there will be a strong tendency for the particles formed to adhere to each other, creating the well-known caking problem. As a related example, the size of crystals decreases if a saturated salt solution is cooled if a surfactant is added (thus reducing the crystal-solution interfacial surface tension).

The surface tension of the system can also be changed by performing grinding under a liquid, thus decreasing the interfacial tension. This gives rise to a variety of parameters, since by adding suitable chemicals (electrolytes or surface active agents) one can modify the end product properties. Conversely, the size of crystals formed from a supersaturated solution of a substance is related to the surface tension (at the solid–liquid interface). Thus, to obtain fine crystals one adds a suitable detergent and thus obtains finer crystals than without.

A typical example is the production of the glass fibers which are used for isolation. In order to keep the negatively charged glass fibers from strong adhesion one sprays a cationic-charged surface active agent, which enhances the isolation by keeping the fibers from compact structure formation. In the past few decades, a specific kind of colloidal system based on *monodisperse* size has been developed for various industrial applications. A variety of metal oxides and hydroxides and polymer lattices have been produced. Monodisperse systems are obviously preferred, since their properties will be easily predicted. On the other hand, *polydisperse* systems will exhibit varying characteristics depending on the degree of polydispersity.

7.5 APPLICATIONS OF COLLOID SYSTEMS

The degree of colloid suspensions stability is an important factor in everyday experience. There are a large number of industrial products where these systems are the basic building blocks, for example, food colloids (e.g., milk, mayonnaise), pollution control (e.g., wastewater, air, oil spill), emulsions, and wastewater treatment. The food colloids are one important example. Take mayonnaise, which is a mixture of vegetable oil, egg yolk, and vinegar, an emulsion of oil-in-water.

The electrostatic forces in many systems play a dominant role, such as the separation process (filtration) and wastewater treatment.

7.5.1 Wastewater Treatment and Control (Zeta Potential)

Wastewater contains different kinds of pollutants (e.g., dissolved substances and suspended particles). Wastewater is treated in suitable plants before the processed water is released into the surroundings. Wastewater contains both soluble and insoluble substances that need to be removed. The substances found in wastewater (solutes) are in either a molecular (such as benzene and coloring substances) or ionic form (such as Na^+, Cl^-, Mg^{++}, K^+, and Fe^{++}).

The concentrations are generally given in various units:

Weight/volume, mg L^{-1}; kg m^{-3}
Weight/w, mg kg^{-1}; parts per million (ppm); parts per billion (ppb)
Molarity, moles L^{-1}
Normality, equivalents L^{-1}

Methods needed to treat these pollution systems also depend on the quantitative amounts present. The specific unit used depends on the amounts present. The unit used for trace amounts, such as benzene, is given in parts per million or parts per billion. The hardness of drinking water (mostly Na and Ca-Mg) concentration is given as milligrams per liter (mg L^{-1}). The typical values as found are in the range of less than 10 mg L^{-1} (soft water) or hard water (over 20 mg L^{-1}).

The presence of a net charge at the particle surface gives rise to an asymmetric distribution of ions in the surrounding region. This means that the concentration of counterions close to the surface is higher than the ions with the same charge as the particle. Thus an electrical double layer is measured around such particles placed in water.

The solids can be removed by filtration and precipitation methods. The precipitation (of charged particles) is controlled by making the particles flocculate by controlling the pH and ionic strength. The latter gives rise to a decrease in charge–charge repulsion, and thus can lead to precipitation and removal of finely divided suspended solids. It is thus found that the most important factor that affects zeta potential is pH. Therefore all zeta potential data must mention its pH. Imagine a particle in suspension with a negative zeta potential. If more alkali is added to this suspension then the particle will exhibit an increase in negative charge. On the other hand, if acid is added to the colloidal suspension, then the particle will acquire an increasing positive charge. During this process, the particle will undergo a change from negative charge to zero charge (where the number of positive charge is equal to negative charge [*point-of-zero-charge*, PZC]). In other words, one can control the magnitude and sign of the surface charge by a potential determining ion. The stability is dependent on the magnitude of electrostatic potential at the surface of the colloid, ψ_o. The magnitude of ψ_o is estimated by using the microelectrophoresis method. When an electric field is applied across an electrolyte, charged particles suspended in the electrolyte are attracted toward the electrode of opposite charge. Viscous forces acting on the particles tend to oppose this movement. When equilibrium is reached between these two opposing forces, the particles move with constant velocity. In this technique, the movement (or rather the speed) of a particle is observed under a microscope when subjected to a given electric field. The field is related to the applied voltage, V, divided by the distance between the electrodes (in centimeters). The velocity is dependent on the strength of electric field or voltage gradient, the dielectric constant of the medium, the viscosity, and zeta potential.

Commercially available electrophoresis instruments are used where the quartz cells designed for any specific system are available. The magnitude of zeta potential, ζ, is obtained from the following relation:

$$\zeta = \mu\ \eta/\varepsilon_o D \tag{7.28}$$

where η is the viscosity of the solution, ε_o is the permittivity of the free space, and D is the dielectric constant. The velocity of a particle in a unit electric field is related to its electric mobility.

In another application, the magnitude of zeta potential is measured as a function of added counterions. The variation in zeta potential has been related to the stability of the colloidal suspension.

The results of a gold colloidal suspension (gold sol) are reported as follows:

Counterion	Velocity	Stability (Flocculation Character)
0 Al+3	3(–)	Very high stability
20 10^{-6} mole	2.(–)	Flocculate (4 hr)
30 10^{-6} mole	0 (zero)	Flocculates fast
40 10^{-6} mole	0.2(+)	Flocculate (4 hr)
70 10^{-6} mole	1(+)	Flocculates slowly

These data show that the charge on the colloidal particles changes from negative to zero (when the particles do not show any movement) to positive, at high counterion concentration. This is a very general picture. Therefore, in wastewater treatment plants one adds counterions until the movement is almost zero and thus one can achieve fast flocculation of pollutant particles. The variation of ζ of silica particles has been investigated as a function of pH. The dissociation of the surface groups -Si(OH) is involved. Under these operations, the zeta potential is constantly monitored by using a suitable instrument. In some plants this is carried out under continuous measurements.

In some biological systems, the charge–charge interactions between larger protein molecules, such as hemoglobin (mol. wt. 68000) the aggregation becomes critical if surface charges change. In biology, the charge interactions are known to play a very important role. There is an extensive number of studies that support this finding in the literature.

7.5.2 Steric Stabilization of Solid or Liquid Colloids

The stability of solid particles or liquid drops can be also controlled by using large molecules (*polymers*). The addition of polymers will result in adsorption on solid or penetration in liquid (Figure 7.10).

The mechanism of polymer stabilization is manyfold. This introduces to the system the following new parameters:

- The colloid is imparted a different charge depending on the polymer. In fact, if the polymer is neutral then the colloid may even become neutral.

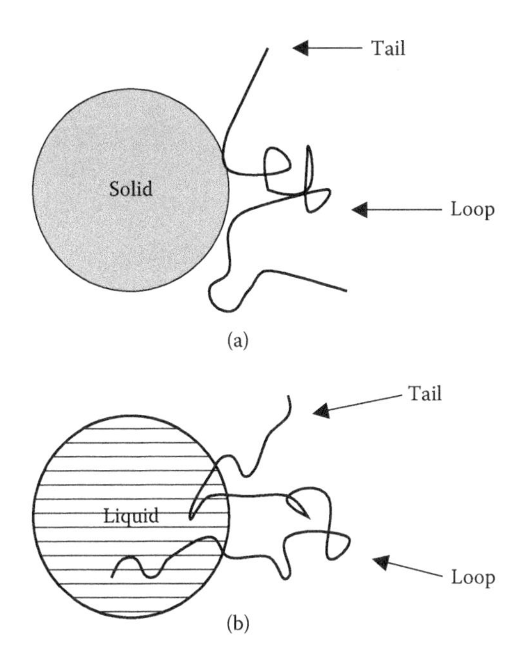

FIGURE 7.10 The state of a polymer at a (a) solid or (b) liquid drop. The polymer may be in three typical states: tail, loop, or coil (helix-coil).

- The size of polymer imparts some special stability.
- The polymer effects the viscosity of the media.

The main criteria for using polymers for colloidal stability are reported as:

- As two colloids approach each other the adsorbed polymers increase repulsion.
- If the polymers interact the two colloids will exhibit increased repulsion as the solvent molecules would push toward this area.
- The polymers after adsorption change in conformation and thus may increase stability.

However, a suitable polymer for a given colloidal system needs to be found. The use of polymers in colloidal suspensions has been a very extensive area of application. A typical example as found in biology is the stabilization of milk.

In various industrial production, materials need to be treated with charged colloidal particles. In such systems, the value of zeta potential analyses need to control the production. In oil drilling, clay colloidal suspensions are used. The zeta potential is controlled so as to avoid clogging the pumping process in the oil well.

It has been found that, for instance, the viscosity of a clay suspension shows a minimum when the zeta potential is changed (with the help of pH from 1 to 7) from 15 to 35 mV. Similar observations have been reported in coal slurry viscosity. The viscosity was controlled by zeta potential.

7.5.3 Industrial Applications of Colloids

Colloids are found to play an important role in many aspects of industrial products. Therefore, there is much research devoted to the following aspects of colloids:

- Production of well-defined colloidal products
- Stability criteria of colloidal products
- Disability criteria
- New applications of novel colloidal preparations

It is useful to mention some industrial products where the aforementioned characteristics are of primary importance: paints and inks, food products, and pharmaceuticals.

The potential differences at different phase boundaries as mentioned earlier are found to have many industrial applications. The application of electrophoresis to the separation and purification of protein is one example. Both electrophoresis and electroosmosis have also attained a certain amount of use in industrial processes.

Electrophoresis techniques are often applied in the purification of natural colloids. For example, they may be separated from uncharged or oppositely charged impurities by electrophc deposited on an electrode. The electrophoretic purification of clay and kaolin are examples. The clay is made into a negatively charged sludge, whose charge may be increased with the addition of small quantities of a base (potential-determining OH ions), so that the suspension becomes a stable colloid while the impurities settle out. The suspension is then subjected to electrophon, thus the colloidal particles are deposited on a rotating metallic cylinder that serves as an anode in which it can be scraped off continuously. The clay is dehydrated during the production.

The colloidal particles may often be deposited on metallic electrodes in the form of adsorbed coatings. Rubber and graphite coatings can be formed in this way, using solvent mixtures (water–acetone) as the dispersion media. The advantage of this method is that additives can firmly be codeposited with, for example, a rubber latex. Thermionic emitters for radio valves are produced in a similar manner. The colloidal suspensions of alkaline earth carbonates are deposited electrophoretically on the electrode, and are later converted to oxides by using an ignition process.

8 Gas Bubbles: Thin Liquid Films and Foams

8.1 INTRODUCTION

If one shakes pure water, no bubbles are observed at the surface. On the other hand, if one shakes a soap solution, bubbles appear at the surface after shaking. Bubbles are also observed on the shores of lakes. This shows that some kind of surface thin liquid film is present in the soap solution. The formation and structure of *thin liquid films* (TLF; such as in foams or bubbles) is the most fascinating phenomena that mankind has studied over many decades. Gas bubbles have been investigated by scientists in a large number of applications in everyday life. It may be accepted that this structure is the closest one comes to observing molecules by the naked eye. TLF is thus the thinnest object one can see without the aid of any kind of microscope. One of the most commonly known thin liquid film structures is the soap bubble or bubbles formed in detergent solutions (such as in dishwashing solutions). Everyone has enjoyed the formation of soap bubbles and the view of the rainbow colors. It may look as if the bubble formation and stability does not have such a great consequence, but in everyday life, bubbles play an important role (for example, from lung function to beer and champagne). In this chapter, the formation and stability of bubbles will be described. Further, even though one cannot see or observe the surface layer of a liquid directly, thin liquid films allow one to make some observations that provide much useful information (Adamson and Gast, 1997; Birdi, 2002, 2009, 2010a, 2010b).

8.2 SOAP BUBBLES AND FOAMS

Let us consider two systems: pure water and a soap (detergent) solution. If pure water is shaken then no bubbles are observed at the surface. All pure organic fluids exhibit no bubble formation on shaking. This means that as an air bubble rises to the surface of the liquid it merely exits into the air. On the other hand, if an aqueous detergent (surface active substance) solution is shaken or an air bubble is created under the surface, then a bubble is formed (Figure 8.1).

Air bubble formation can be described as follows:

- Air bubble inside liquid phase, at the surface the bubble detaches and moves up under gravity.

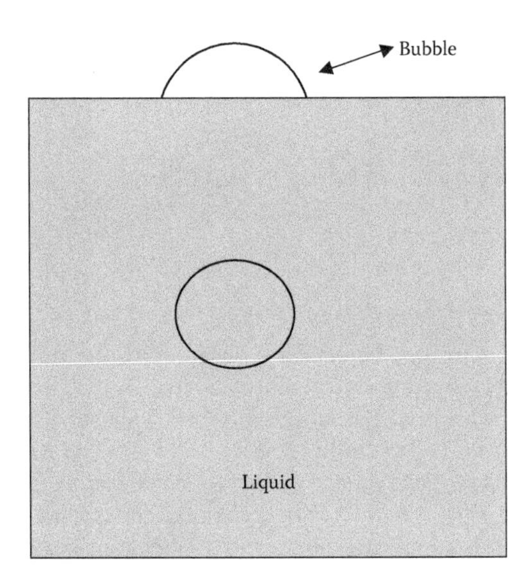

FIGURE 8.1 Formation of a bubble.

- The detergent molecule forms a bilayer in the bubble film. The water in between is the same as the bulk solution. This may be depicted as follows:
  - Surface layer of detergent
  - Bubble with air and a layer of detergent
- Bubble at the surface forms double layer of detergent with some water in between (thin liquid film, varying from 10 μm to 100 μm).

Even very minute amounts (around parts per million) of surface active substances give rise to stable bubbles. This method has been used to detect the presence of surface active substances in water. A bubble is composed of a TLF with two surfaces, each with a polar end pointed inward and the hydrocarbon chains pointing outward (Figure 8.2). The water inside the films will move away (due to gravity and evaporation) giving rise to the thinning of the film. Since the thickness approaches the dimensions of the light wavelength, one observes varying interference colors.

The reflected ray will interfere with the incident wavelength. The consequence of this will be that depending on the thickness of the film, colors will appear. The most amazing observation is that the thickness of these films is comparable to the wavelength of light, especially when the thickness of the film is approximately the same as the wavelength of the light (i.e., between 400 and 1000 Å). The black film is observed when the thickness is the same as the wavelength of the light (approximately 500 to 700 Å). Thus, this provides the closest visual observation of two molecule thick film by eyesight.

8.2.1 Application of Bubbles in Technology

One of the most important roles of bubbles is in the food industry (such as ice cream, champagne, and beer). The stability and size of the bubbles determines the taste and

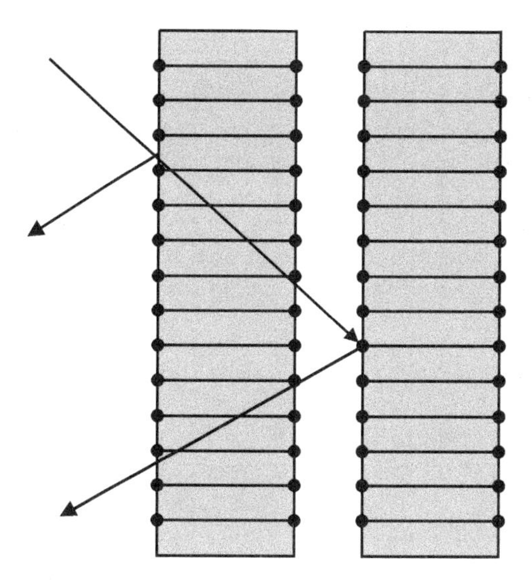

FIGURE 8.2 Thin liquid film and reflection of light.

the looks of the product. Especially, in the case of champagne, both the size and the stability of bubbles have been found to determine the impact of taste and flavor. It has been estimated that in a bottle of volume 750 mL, there will be about 50 million bubbles (if average radius is 0.1 mm). This is a very rough estimate. However, more accurate estimates have been made by using photography of bubbles. In this industry much research has been made on the determination of the factors that control the bubble formation and stability. Another example is in ice creams where air bubbles are trapped in the frozen material.

8.2.2 Foam Formation (Thin Liquid Films)

Ordinary foams from detergent solutions are thick initially (micrometer), and as fluid flows away due to gravity or capillary forces or surface evaporation, the film becomes thinner (few hundred angstroms). Foams are an essential part in many processes, both in industry and biology.

Foam consists of:

- Air on one side
- Outer monolayer of detergent molecule
- Some amount of water
- Inner monolayer of detergent molecule
- Air on outer side

This can be depicted (schematically) as follows:

```
DETERGENT MOLECULEWATERDETERGENT MOLECULE
DETERGENT MOLECULEWATERDETERGENT MOLECULE DETERGENT
MOLECULEWATERDETERGENT MOLECULE DETERGENT
MOLECULEWATERDETERGENT MOLECULE
```

The orientation of detergent molecule in TLF is such that the polar group (OO) is pointing toward the water phase and the apolar alkyl part (CCCCCCCCCC) is pointing toward the air.

```
AirCCCCCCCCCCOOWATERWATEROOCCCCCCCCCCAir
AirCCCCCCCCCCOOWATERWATEROOCCCCCCCCCCAir
AirCCCCCCCCCCOOWATERWATEROOCCCCCCCCCCAir
AirCCCCCCCCCCOOWATERWATEROOCCCCCCCCCCAir
AirCCCCCCCCCCOOWATERWATEROOCCCCCCCCCCAir
AirCCCCCCCCCCOOWATERWATEROOCCCCCCCCCCAir
```

The thickness of the water phase can vary from over 100 μm to less than 100 nm. Foams are thermodynamically unstable, since there is a decrease in total free energy when they collapse. As the thickness comes around the wavelength of light (nm), one starts to observe rainbow colors (arising from interference). The TLF is even smaller in thickness (50 Å or 5 nm) (Figure 8.3).

However, certain kinds of foams are known to persist for very long periods of time and many attempts have been made to explain their metastability. TLF may be regarded as a kind of condenser. The repulsion between the two surfactant layers

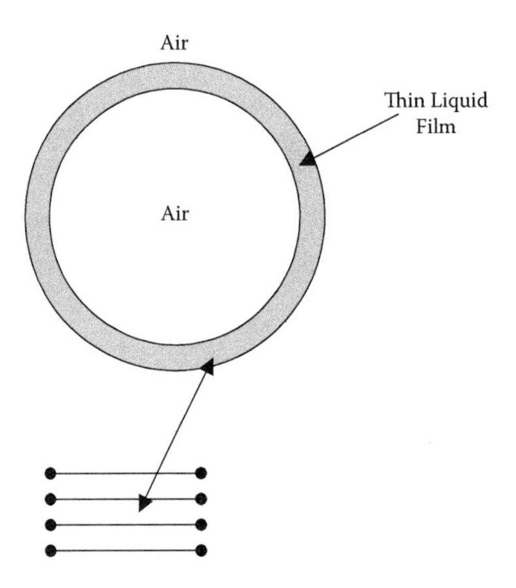

FIGURE 8.3 Thin liquid film structure. (Schematic drawing of the structure of a black thin film showing the aqueous phase and the detergent molecules.)

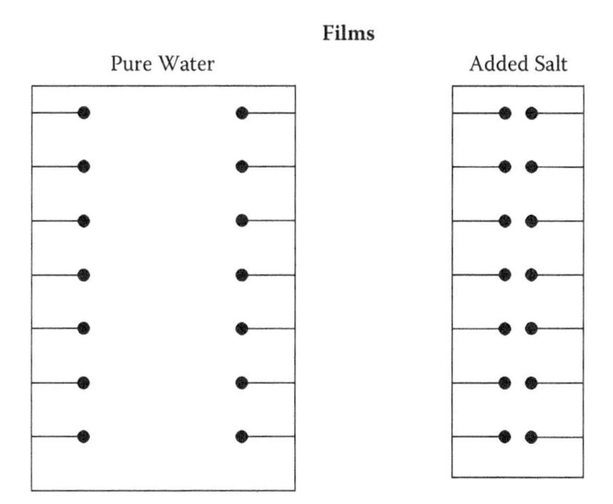

FIGURE 8.4 Thickness of foam films and added ions.

(Figure 8.4) will be determined by the electrical double layer (EDL). The effect of added ions to the solution is to make the EDL contract and this leads to thin films.

The foam films look black-gray when the thickness is around 50 Å (5 nm), which is almost the size of the bilayer structure of the detergent (i.e., twice the length [ca. 25 Å] of a typical detergent molecule). Actually, it is remarkable that one can see the two molecule thin structure. The rainbow colors are observed since the light is reflected by the varying thickness of the TLF of the bubble.

It may not be obvious at first sight, but in the beer industry the foaming behavior is one of the most important characteristics. Beer in a bottle is produced under high pressure of CO_2 gas. As soon as one opens a beer bottle the pressure drops and the gas, CO_2, is released, which gives rise to foaming. In common behavior the foam stays inside the bottle. The foaming is caused due to the presence of different amphiphilic molecules (e.g., fatty acids, lipids, and proteins). This foam is very rich as the liquid film is very thick and contains lots of aqueous phase (such foams are called *kúgelschaum*). The foam fills the empty space in the bottle and under normal conditions it barely spills out. However, under some abnormal conditions the foam is highly stable and starts to pour out of the bottle and is considered undesirable. In some reports one has found that addition of heavy metal ions could change the foaming characteristics (Birdi, 1989).

With regard to foam stability, it has been recognized that the surface tension under film deformation must always change in such a way as to resist the deforming forces. Thus, tension in the film where expansion takes place will increase, while it will decrease in the part where contraction takes place. There is, therefore, a force tending to restore the original condition. The film is elastic, thus the term *elasticity* has been defined as:

$$E_{film} = 2\ A\ (d\ \gamma/dA) \tag{8.1}$$

where E_{film} and A are the elasticity and area of the film, respectively, and γ is the surface tension of the surface deformed. One of the most important applications is the bubble formation in champagne. The size and the number of bubbles are important for the impact of taste. The stability also has much impact on the looks and taste as well. Taste is related to the size and number of bubbles, and as well as how long the bubbles are stable. The stability of any foam film is related to the kinetics of thinning of the thin liquid film. As the thickness reaches a critical value, the stability becomes critical. It was recognized by Gibbs that an unstable state will conform when the film diverges from bulk system properties. This thickness was mentioned to be in the range of 50 to 150 Å. This state is called the *black* film and the random motion of the molecules may easily give rise to a rupture of the thin liquid film. The flow of liquid is determined by gravity forces. It was found that assuming the fluid in film has the same viscosity and density, then the mean velocity will not exceed 1000 D_{film}^2, where the latter is distance between the lamella:

D (mm)	Flow (mm/sec)
0.01	0.1
0.001	0.001

8.2.3 Criteria of Foam Stability

As it is known, if one blows air bubbles in pure water no foam is formed. On the other hand, if a detergent or protein (amphiphile) is present in the system, adsorbed surfactant molecules at the interface give rise to foam or soap bubble formation. Foam can be characterized as a coarse dispersion of a gas in a liquid, where gas is the major phase volume. The foam, or the lamina of liquid, will tend to contract due to its surface tension, and a low surface tension would thus be expected to be a necessary requirement for good foam-forming property. Furthermore, in order to be able to stabilize the lamina it should be able to maintain slight differences of tension in its different regions. It is therefore clear that a pure liquid, which has constant surface tension, cannot meet this requirement. The stability of such foams or bubbles has been related to monomolecular film structures and stability. For instance, foam stability has been shown to be related to the surface elasticity or surface viscosity, η_s, in addition to other interfacial forces. Studies have shown that foam destabilization is related to the packing and orientation of mixed films, which can be determined from monolayer studies. Since very small (microgram) amounts are needed for such studies, the monolayer method has been useful in a large variety of system studies. It is also worth mentioning that foam formation from monolayers of amphiphiles constitutes the most fundamental process in everyday life. The other assemblies, such as vesicles and bilayer lipid membranes (BLMs) are somewhat more complicated systems, which are also in equilibrium with monolayers. It is important to mention that foam does not form in organic liquids, such as methanol and ethanol.

Although the surface potential, φ, the electrical potential due to the charge on the monolayers, will clearly affect the actual pressure required to thin the lamella to any given thickness, we shall assume for the purpose of a simple illustration that $1/k$, the mean Debye-Huckel thickness of the ionic double layer, will influence the ultimate thickness when the liquid film is under a relatively low pressure. Let us also assume that each ionic atmosphere extends only to a distance $3/k$ into the liquid when the film is under a relatively low excess pressure from the gas in the bubbles. This value corresponds to a repulsion potential of only a few millivolts. Thus at about 1 atm pressure:

$$h_{film} = 6/k + 2(\text{monolayer thickness}) \tag{8.2}$$

For charged monolayers adsorbed from 10^{-3} n-sodium oleate, the final total thickness, h_{film}, of the aqueous layer should thus be of the order 600 Å (i.e., $6/k$ or 18 Å). To this value, 60 Å (60 10^{-10} m) needs to be added for the two films of oriented soap molecules, giving a total of 660 Å. The experimental value is 700 Å. The thickness decreases on the addition of electrolytes, as also suggested by the preceding equation. For instance, the value of h_{film} is 120 Å in the case of 0.1 M-NaCl. The addition of a small amount of certain nonionic surface active agents (e.g., n-lauryl alcohol, n-decyl glycerol ether, laurylethanolamide, and laurylsufanoylamide) to anionic detergent solutions has been found to stabilize the foam. It has been suggested that the mode of packing is analogous to the palisade layers of the micelles and the surface layers of the foam lamellae. Measurements have been carried out on the excess tensions, equilibrium thicknesses, and compositions of aqueous foam films stabilized by either n-decyl methyl sulfoxide or n-decyl trimethyl ammonium-decyl sulfate and containing inorganic electrolytes. It was recognized at a very early stage (Birdi, 2002, 2007) that the stability of a liquid film must be greatest if the surface pressure strongly resists deforming forces.

It has been shown (Birdi, 2002, 2009; Friberg et al., 2003) that a correlation exists between foam stability and the elasticity of the film, that is, a monolayer. In order for elasticity to be large, surface excess must be large. Maximum foam stability has been reported in systems with fatty acid and alcohol concentrations well below the minimum in γ. Similar conclusions have been observed with n-$C_{12}H_{25}SO_4Na$ (SDS) + n-$C_{12}H_{25}OH$ systems, which give minimum in γ versus concentration with maximum foam at the minimum point (Chattoraj and Birdi, 1984) due to mixed monolayer formation. It has been found that SDS + $C_{12}H_{25}OH$ (and some other additives) make *liquid crystalline* structures at the surface. This leads to a stable foam (and liquid-crystalline structures). In fact, small amounts of $C_{12}H_{25}OH$ (less than 10%) are used as foam enhancers in these systems. The foam drainage, surface viscosity, and bubble size distributions have been reported for different systems consisting of detergents and proteins. Foam drainage has been investigated by using an incident light interference microscope technique. The foaming of protein solution is of theoretical interest and also has wide applications in the food industry (Friberg et al., 2003). Further, in the fermentation industry where foaming is undesirable, foam is generally caused by proteins. Since, mechanical defoaming is expensive due to the high power required, antifoam agents are generally used. On the other hand, antifoam agents are not desirable in some of these systems, as for instance, in food products. In addition, the antifoam agents deteriorate the gas dispersion due to increased coalescence of the bubbles. Foams are stabilized by proteins and these are dependent on pH and electrolytes.

The high foaming capacity is explained by the stability of the gas–liquid interface due to the denaturation of protein, especially due to their strong adsorption at the interface, which gives rise to the stable monomolecular films at the interface. The foam stability is caused by the film cohesion and elasticity. Further, the degree of foaming of a BSA (bovine serum albumin) aqueous solution has been investigated. The effect of electrolytes and alcohol was investigated. A good correlation was found between the adsorption kinetics and the foaming properties. The effect of partial denaturation on the surface properties of ovalbumin and lysozyme has been reported. Most protein molecules exhibit increased hydrophobicity at the interface as denaturation proceeds, due to the exposure to the outer surface of the buried hydrophobic residues (in the native state). The hydrophobicity of proteins (as described in Chapter 5) has been found to give a fairly good correlation to emulsion stability in food proteins. The surface tension of these proteins decreased greatly as denaturation proceeded. The emulsifying and foaming properties of proteins were remarkably improved by heat denaturation without coagulation. The emulsifying properties increased and were found to exhibit a correlation with surface hydrophobicity. Protein foaming properties increased with denaturation. The foaming power and foam stability of $C_{12}H_{25}SO_4Na$ (SDS)–ovalbumin complexes did not improve as much as with heat-denatured protein. The surface hydrophobicity showed an increase. It is thus safe to conclude that the heat and detergent denatured proteins are unfolded by different mechanisms. These studies are in accord with the unfolding studies carried out comparing urea or SDS unfolding, by fluorescence.

Foam structure—The foam as thin liquid film has a very fascinating structure. If two bubbles of the same radius come into contact with each other, this leads to the formation of contact area and subsequently to formation of one large bubble. This leads to the following stages:

1. Two bubbles of same radius
2. Two bubbles touch each other and form a contact area
3. Formation of only one bubble

In stage 2 the energy of the system is higher than in stage 1, since the system has formed a contact area (dA_c). The energy difference between stage 2 and 1 is $\gamma\, dA_c$. When final stage is reached, stage 3, there will be a decrease in total area by 41% (i.e., the sum of the area of two bubbles is larger than that of one bubble). This means that stage 3 is at a lower energy state than the initial state I ($\gamma\, d\, A_{II-III}$).

When three bubbles come into contact the equilibrium angle will be 120°. The angle of contact relates to the systems equilibrium state, which is 120° from simple geometrical considerations. If four bubbles are attached to one another then the angle at equilibrium is 109°28° (Figure 8.5).

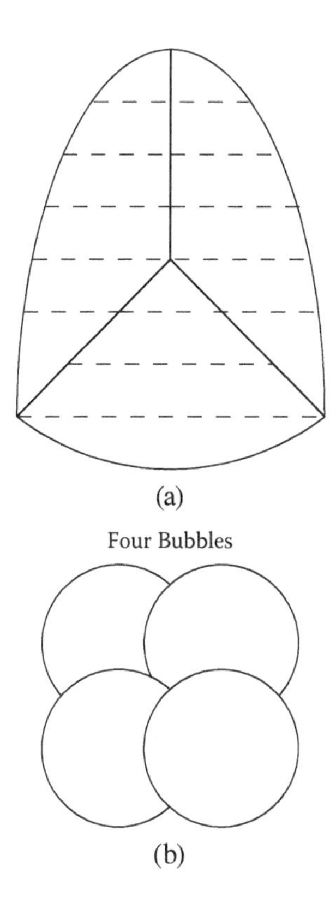

FIGURE 8.5 Foam structure of (a) three-bubble and (b) four-bubble aggregates.

8.2.3.1 Foam Formation of Beer and Surface Viscosity

The surface and bulk viscosities not only reduce the draining rate of the lamella, but also help in restoration against mechanical, thermal, or chemical shocks. The highest foam stability is associated with appreciable η_s and yield value.

The overfoaming characteristic of beer (gushing) has been the subject of many investigations. The extreme case of gushing is when a beer upon opening starts to foam out of the bottle, and in some cases empties the whole bottle. The relationship between surface viscosity and gushing was reported by various investigators. The various factors were described for the gushing process: pH, temperature, and metal ions, which could lead to protein denaturation.

The stability of a gas (i.e., N_2, CO_2, and air) bubble in a solution depends on its dimensions. A bubble with a radius greater than a *critical magnitude* will continue to expand indefinitely and degassing of the solution would take place. Bubbles with a radius equal to the critical value would be in equilibrium, while bubbles with radius less than the critical value would be able to redissolve in the bulk liquid. The magnitude of the *critical radius*, R_{cr}, varies with the degree of saturation of the liquid, that is, the higher the level of supersaturation the smaller the critical radius. It has been suggested that there is nothing unusual in the stability of the beer and, although carbon dioxide is far from an ideal gas, empirical work supports this conclusion. A possible connection between surface viscosity and gushing has been reported. Nickel ion, a potent inducer of gushing, has been reported to give rise to a large increase in the surface viscosity of beer. Other additives, in addition to Ni, such as Fe or humulinic acid, which cause gushing, have also been reported to result in large increases in surface viscosity. On the other hand, additives that are reported to inhibit gushing, such as EDTA (ethylenediaminetetraacetic acid, a chelating agent), have been reported to decrease surface viscosity of beer. This relation between surface viscosity and gushing has been suggestive that an efficient gushing inhibitor should be very surface active in order to be able to compete with gushing promoters, but incapable of forming rigid surface layers (i.e., high surface viscosity). Unsaturated fatty acids, such as linoleic acid, are potent gushing inhibitors, since they destabilize the surface films. The surface viscosity was investigated by the oscillating-disc method. It has been found that low surface viscosity (0.03 to 0.08 g/sec) beer surfaces give nongushing behavior. Beers with high surface viscosity (2.3 to 9.0 g/sec) gave gushing.

8.2.4 Antifoaming Agents (Destabilizing Foam Bubbles)

In many cases, foaming is undesirable (such as in machine dishwashing and wastewater treatment). The main criteria for antifoaming molecules is that these exhibit the following characteristics (Birdi, 2009):

- Do not form mixed monolayers
- Reduce surface viscosity (thus destabilizing the foam films)
- Low boiling point liquid additives (such as ethanol).

Defoaming agents are commercially available and designed for any particular system requirement.

8.3 APPLICATIONS OF FOAMS AND BUBBLES

Foams and bubbles are easily created and require very little energy input. This technology, therefore, makes it very useful in those applications where bubbles can be of positive benefit.

8.3.1 Water Purification Technology

The biggest challenge mankind is facing is the need to supply pure drinking water worldwide. The world population increase (from 1900 to 2000 by a factor of 4) is much faster than the availability of clean drinking water supplies. In addition, the increased need for water in industrial production also adds further burden on the water supply. The purification of water for households has been developed during the past decades. Pollutants as found in wastewater are different in origin and concentration. Solid particles are mostly removed by filtration, but the colloidal particles are not easily removed by this method. Solute compounds are rather difficult to remove, especially toxic substances with very low concentration. *Flotation* has been used with great advantage in some cases where sedimentation cannot remove all the suspended particles. Following are some examples where flotation is being used with much success:

- Paper fiber removal in the pulp and paper industry
- Oils, greases, and other fats in food, oil refinery, and laundry wastes
- Clarification of chemically treated waters in potable water production
- Sewage sludge treatment

Many of the industrial wastewaters amenable to clarification by flotation are colloidal in nature, for example, oil emulsions, pulp and paper wastes, and food processing. For the best results, such wastes must be coagulated prior to flotation. In fact, flotation is always the last step in the treatment. In order to aid the flotation effectivity, surfactants are used. This leads to lower surface tension and foaming. The latter helps in retaining the particles in the foam under flotation. Further, the effectiveness of flotation is also dependent on what kind of gas is used to make the bubbles (e.g., air and CO_2).

8.3.2 Bubble Foam Purification of Water

It was found many decades ago that foam or bubbles could be used to purify wastewater. A simple device as that used for laboratory froth flotation studies is shown in Figure 8.6.

The bubbles are formed in the sintered glass as air or other suitable gas (N_2, CO_2, etc.) is bubbled through the solution containing the solid suspension. Suitable flotation agent (a suitable surface active agent) is added and the air is bubbled. Surface active pollutants in wastewater have been removed by bubble film separation

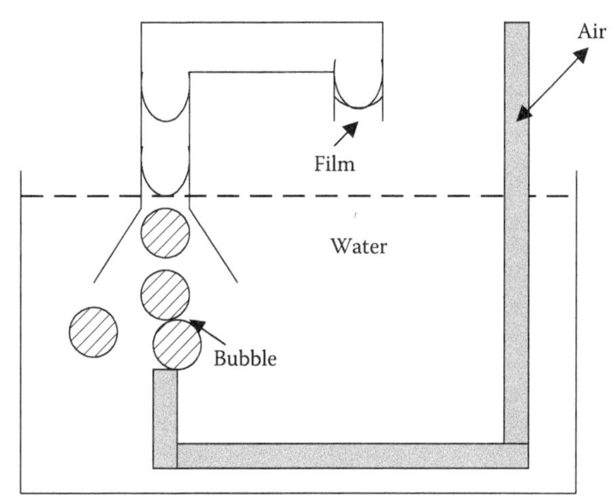

FIGURE 8.6 Bubble foam separation method for wastewater purification.

methods. Especially, very minute concentrations are easily removed by this method, which is more economical than more complicated methods (such as active charcoal, filtration, and other chemical methods). This method is now commercially available for small systems such as fish tanks (Birdi, 2009, 2010a). The principle in this procedure is to create bubbles in wastewater tanks and to collect the bubble foam at the top (Figure 8.6).

Bubbles are blown into the inverted funnel. Inside the funnel, the bubble film is transported away and collected. Since the bubble film consists of surface active substances and water (and salts, etc.), it is thus seen that even minute amounts (less than a milligram per liter) of surface active substances will accumulate at the bubble surface. As previously shown, it would require a large number of bubbles to remove a gram of substance. However, since one can blow thousands of bubbles in a very short time, the method is very feasible.

9 Emulsions, Microemulsions, and Lyotropic Liquid Crystals

9.1 INTRODUCTION

One of the most important aspects of surface and colloid chemistry is the subject of oil–water emulsion technology. Oil is sparingly soluble in water and vice versa. In this chapter special applications of surface chemistry principles pertaining to oil and water phases will be considered. The science of emulsion is also important since many biological systems use the same principles. The essential subject is based on the fact that oil and water *do not mix* if shaken.

As is well known if one shakes oil and water, oil breaks up into small drops (about few millimeters in diameter) but these drops join together rather quickly to return to their original state:

1. Oil phase and water phase
2. Mixing
3. Oil drops in water phase
4. After short time
5. Oil phase and water phase (same as beginning)

However, one finds that oil and water can be dispersed with the help of suitable *emulsifiers* (surfactants) to give *emulsions* (Becher, 2001; Friberg et al., 2003; Birdi, 2008, 2010a; Sjoblom et al., 2008). The basic reason is that the interfacial tension (IFT) between oil and water is around 50 mN/m, which is high and leads to formation of large oil drops. On the other hand, the addition of suitable emulsifiers can reduce the value of IFT to very low values (even much less than 1 mN/m). Emulsion formation means that oil drops remain dispersed for a given length of time (up to many years). The stability and the characteristics of these emulsions are related to the area of applications. Emulsions are a mixture of two (or more) immiscible substances. Some everyday common examples include milk, butter (fats, water, salts), margarine, mayonnaise, and skin creams. In butter and margarine, the continuous phase consists of lipids. These lipids surround the water droplets (water-in-oil emulsion).

All technical emulsions are prepared by some convenient kind of mechanical agitation or mixing. Remarkably, the natural product, milk, is made by the organism without any agitation but inside the glands. Emulsions are one of the most important application areas of surface active compounds.

These oil–water systems are generally described as three different kinds:

- Emulsions
- Microemulsions
- Liquid crystals (LC) (and lyotropic LC)

The different kinds of emulsions will be described separately. Emulsions are systems where water and oil both need to be applied to an application. This may be skin treatment or shoe shine. In other words, both of these two components (water and oil which do not mix) can be applied simultaneously. This also allows functions to be performed that are dependent on the properties of water or oil. In most emulsion systems, two immiscible liquids, such as the water and oil phases, are involved. Information about the IFT as well as the solubility characteristics of *surface active substances* (SAS) is needed to stabilize the emulsions.

Microemulsions are microstructured mixtures of oil, water, emulsifiers, and other substances. Since microemulsions differ in many ways from the ordinary emulsion structure, the former will be described separately. Liquid crystals (LCs) are substances that exhibit special melting characteristics. Further, some mixtures of surfactant, water, and cosurfactant may also exhibit lyotropic LC properties. The emulsion technology is basically thus concerned in preparing mixtures of two immiscible substances: oil and water by adding suitable surface active agents (emulgators, cosurfactants, and polymers). The emulsion technology is thus very varied, since there are many simple systems (such as skin creams) and there are also very complex systems (such as milk).

9.2 FORMATION OF EMULSIONS (OIL AND WATER)

When a surface active substance is added to a system of oil–water, the magnitude of IFT decreases from 50 mN/m to 30 mN/m (or lower, less than 1 mN/m). This leads to the observation that on shaking an oil–water system the decreased IFT leads to smaller drops of the dispersed phase (oil or water). The smaller drops also lead to a more stable emulsion. Depending on the surfactant used, one will obtain an oil-in-water (O/W) or water-in-oil (W/O) emulsion. These experiments where an oil–water or oil–water–surfactant are shaken together are shown in Figure 9.1.

These emulsions are all opaque since they reflect light. Some typical oil–water IFT values are given in Table 9.1. These data show certain trends. The decrease in IFT is much smaller with a decrease in the alkyl chain in the case of alkanes than alcohols.

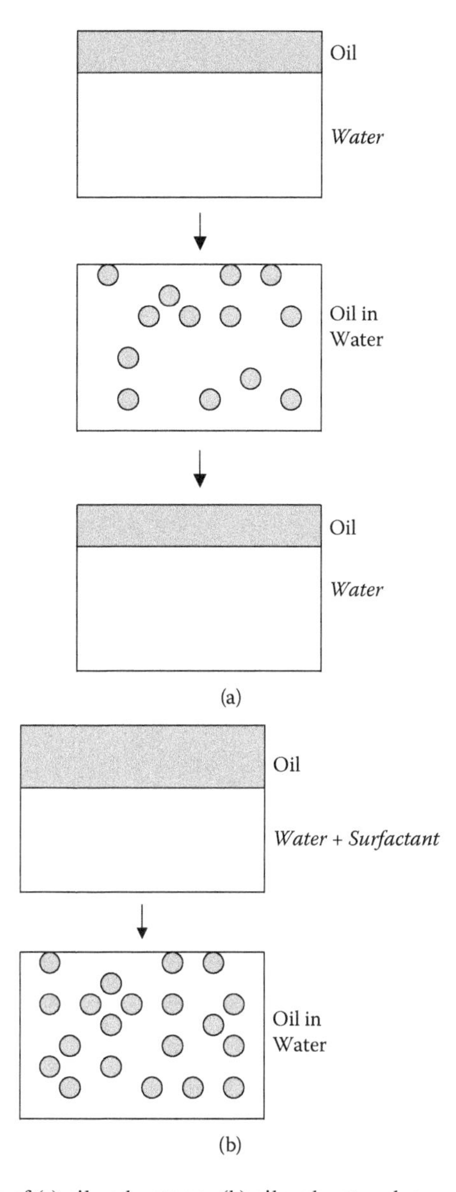

FIGURE 9.1 Mixing of (a) oil and water or (b) oil and water plus a surfactant by shaking.

9.2.1 Types of Oil–Water Emulsions

Emulsions are some of the most important structures that are prepared specifically for a given application. For example, a day cream (skin cream) has different characteristics and ingredients than a night cream. One of the main differences in emulsions is whether oil droplets are dispersed in the water phase or the water drops are dispersed in the oil phase. This can be determined by measuring the conductivity,

TABLE 9.1
Magnitudes of Interfacial Tensions of Different Organic Liquids against Water (20°C)

Oil Phase	IFT (mN/m)
Hexadecane	52
Tetradecane	52
Dodecane	51
Decane	51
Octane	51
Hexane	51
Benzene	35
Toluene	36
CCl4	45
CCl3	32
Oleic acid	16
Octanol	9
Hexanol	7
Butanol	2

since it is higher for O/W than for W/O emulsion. Another useful property is that O/W will dissolve water whereas W/O will not. This thus shows that one will choose W/O or O/W depending on the application area. In the case of skin emulsions, the type of emulsion is of much importance.

9.2.2 Oil-in-Water Emulsions

The main criteria for O/W emulsion will be that if water is added, then it will be miscible with the emulsion. Further, after water evaporates, the oil phase will be left behind. Thus, if the oil phase on the substrate is needed (such as skin, metal, or wood), then an O/W type emulsion should be used.

9.2.3 Water-in-Oil Emulsions

The criteria for W/O emulsion is that it is miscible with oil. That means that if the emulsion is added to some oil then a new but diluted W/O emulsion is obtained. In some skin creams the use of an W/O type emulsion is the preference (especially if one needs an oil-like feeling after its application). On the other hand, if an oil-free surface is needed, then an O/W emulsion is preferred.

9.2.4 Hydrophilic–Lipophilic Balance Values of Emulsifiers

The emulsion technology requires a very exact knowledge of the physicochemical properties of the various components. The lipid components used in emulsions are investigated in mixed film monolayer systems. These studies are useful in the

TABLE 9.2
HLB Values of Different Emulsifiers (Commonly Used in Emulsions)

Emulsifier Solubility in Water	HLB	Application
Low solubility	0–2	W/O
Low solubility	4–8	W/O
Soluble	10–12	Wetting agent
High solubility	14–18	O/W

determination of the feasibility of components. The emulsifiers used exhibit varying solubility in water (or oil) as related to the hydrophilic–lipophilic balance (HLB) value. This will thus have consequences on the emulsion. Let us consider a system where we have oil and water. If we add an emulsifier to this system, then the latter will be distributed in both the oil and water phases. The degree of solubility in each phase will depend on its structure and HLB character. The emulsifiers as used in making emulsions are characterized with regard to the molecular structure. The amphiphile molecules consist of HLB characteristics. Thus, each emulsifier that may be needed for a given system (for example, if an O/W or W/O emulsion is needed), will need a specific HLB value. The data in Table 9.2 provide a rough estimation of the HLB needed for a given system of emulsion. In general, if the emulsifier dissolves in water, then upon adding oil, an O/W emulsion will result. Conversely, if the emulsifier is soluble in oil, then upon adding water, a W/O emulsion will result.

W/O emulsions are formed by using HLB values between 3.6 and 6. This suggests that one generally uses emulgators, which are soluble in the oil phase. O/W emulsions need an HLB around 8 to 18. These HLB criteria are only a very general observation. However, it should be noted that HLB values alone do not determine the emulsion type. Other parameters, such as temperature, properties of oil phase, and electrolytes in aqueous phase also matter. The HLB values have no relation to the degree of emulsion stability. The HLB values of some surface active agents are provided in Table 9.3.

The HLB values decrease as the solubility of surface active agent *decreases in water*. Solubility of cetyl alcohol in water (at 25°C) is less than a milligram per liter mg/liter. It is thus obvious that in any emulsion cetyl alcohol will be present mainly in the oil phase, whereas sodium dodecyl sulfate (SDS) will be mainly found in the water phase. The empirical HLB values are found to have significant use in applications in emulsion technology. It has been shown that the HLB is related, in general, to the distribution coefficient, K_D, of the emulsifier in the oil and water phases:

$$K_D = C(\text{water})/C(\text{oil}) \tag{9.1}$$

where C(water) and C(oil) are the equilibrium molar concentrations of the emulsifier in the water and oil phase, respectively. Based on this definition of K_D, the magnitude of HLB can be estimated as follows:

$$(\text{HLB} - 7) = 0.36 \ln (K_D) \tag{9.2}$$

TABLE 9.3
HLB Values of Some Typical Surface Active Agents

SAA	HLB
Na-lauryl sulfate	40
Na-oleate	18
Tween80 (sorbitan monooleate EO20)	15
Tween81 (sorbitan monooleate EO6)	10
Ca-dodecylbenzene sulfonate	9
Sorbitan monolaurate	9
Soya lecithin	8
Sorbitan monopalmiate	7
Glycerol monolaurate	5
Sorbitan monostearete	5
Span80 (sorbitan monooleate)	4
Glycerol monostearate	4
Glycerol monooleate	3
Sucrose distearate	3
Cetyl alcohol	1
Oleic acid	1

Based on this thermodynamic relation it can be suggested that there is a relation between HLB and emulsion stability and structure. The HLB values can also be estimated from the structural groups of the emulsifier (Table 9.4). Table 9.4 can be useful in those cases where the HLB value needs to be estimated.

There are extensive applications of food emulsifiers. These emulsifiers must satisfy special requirements in order to be useful in the food industry. Toxicity is

TABLE 9.4
HLB Group Numbers

Group	Group Number
Hydrophilic	
-SO_4Na	39
-COOH	21
-COONa	19
Sulfonate	11
Ester	7
-OH	2
Lipophilic	
-CH-	0.5
-CH2	0.5
-CH3	0.5
-CH_2CH_2O-	0.33

determined from animal tests. The test determines the amount of a substance that causes 50% (or more) of the test animals to die (lethal dosage, LD50). Therefore, food emulsions are subject to strict controls (Friberg et al., 2003).

9.2.5 Methods of Emulsion Formation

If one shakes oil and water, the oil breaks up into drops. However, these will quickly coalescence and return to the original state of two different phases. The longer one shakes, the more drops reduce in size. In other words, the mechanical energy put into the system makes the drops smaller in size. Emulsions are made based on different procedures. These can be where mechanical agitation is used. There are other methods that are also used. The emulsion technology is very much state-of-the-art type of industry (Birdi, 2002; Friberg et al., 2003; Sjoblom et al., 2008). Therefore, vast literature exists about methods used for any specific emulsion. In a simple case, an emulsion may be based on three necessary ingredients: water, oil, and emulsifier. In other words, it needs to be determined which weight proportions are needed to mix these substances to obtain a stable (or maximum stability) emulsion at a given temperature. This may be more conveniently carried out in a phase study in the triangle (Figure 9.2). The micellar region exists on the water–surfactant line.

Near the surfactant region is the *crystalline* or *lamellar* phase. This is the region for hand soaps. The ordinary hand soap is mainly salt of fatty acid (typical composition is coconut oil fatty acids or mixtures; 85%) plus water (15%) and perfume, and some salts. X-ray analysis has shown that the crystalline structure consists of a series of layers of soap separated by a water layer (with salts). The hand soap is produced by extruding under high pressure. This process aligns the lamellar crystalline structure lengthwise. If the degree of expansion versus temperature is measured, then the expansion is twice the length than the width. It is further seen that complex structures are present in other regions in the phases (Figure 9.2). The diagram is strongly dependent on temperature, and therefore such studies are carried out at different temperatures.

In practice, what one does is as follows. A suitable number (over 50) test samples are prepared by mixing each component in varying weights to represent a suitable

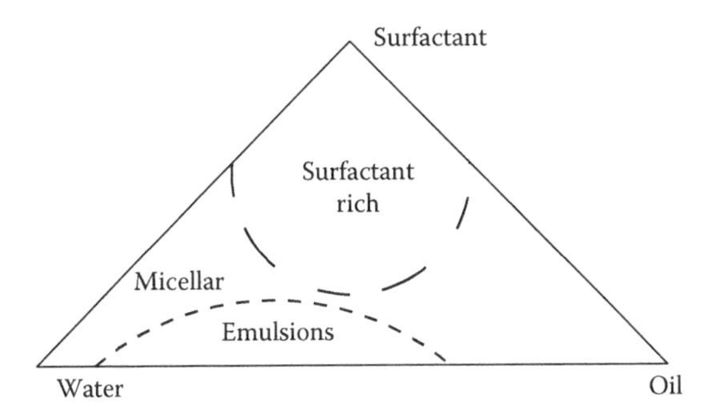

FIGURE 9.2 Different phase equilibriums in a water–surfactant (emulsifier)–oil mixture system.

number of the regions. The test samples are mixed under rotation in a thermostat over a few days to reach equilibrium. The test samples are centrifuged and the phases are analyzed. The phase structure is investigated by using a suitable analytical method. Studies of multicomponent systems such as as these have shown very large numbers of phases. However, by analyzing some typical systems, there are some trends that can be used as guidelines.

Another historically very well investigated system consists of:

- Water
- Potassium caprate (K-caprate)
- n-Octanol

The phases were determined as indicated in Figure 9.3. The system is a very useful example for understanding what phase equilibriums are involved when the three components are mixed. Some characteristics are noticeable in this system, which point out the significance of ratios between KC:O. For example, the aqueous phase region is extrapolated to 1 mol octanol:2 mol K-caprate. This shows that the 1:2 ratio dominates the phase region. It has been reported in other studies, such as monolayers on water films of lipids, that such mixtures are indeed found. The three-phase region is extrapolated to show that 1 mol octanol:1 mol KC is the ratio. In a much simplified description here, in such complicated phase equilibriums some simple molecular ratios indicate the phase boundaries. Thus in general, one may safely conclude that these *molecular ratios* would be useful when working with emulsions. The observation that exact ratios exist between different components at the phase lines suggests that some kind of molecular aggregates are formed. These correspond to the formation of some liquid–crystalline structures. Confirmation of these molecular aggregates has been found from monolayer studies of mixed films spread on water (Chattoraj and Birdi, 1984; Birdi, 1989). A similar conclusion was reached when

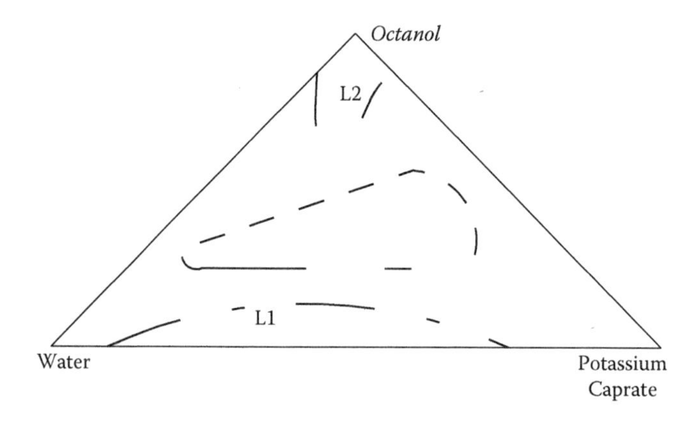

FIGURE 9.3 Phase diagram for the system: K-caprate (PK) + water + n-octanol (22°C). All compositions are given in weight %. L1, micellar phase; L2, reverse micellar phase.

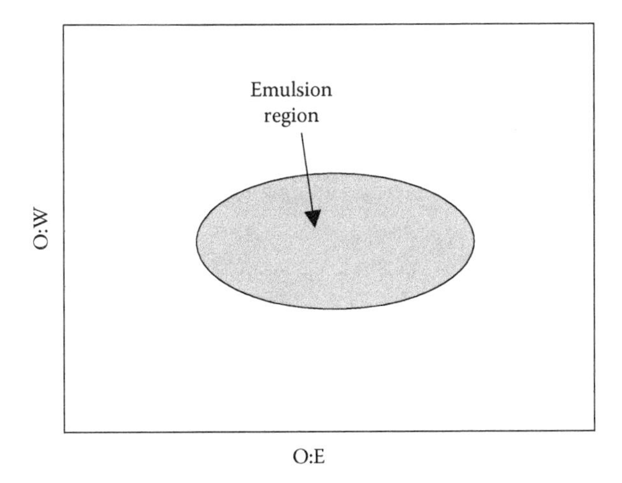

FIGURE 9.4 Emulsion region based on the ratio of oil (O):water (W) versus oil (O):emulsifier (E).

investigating microemulsions (as described later). Further, in practice a given emulsion with some specified range of ratio between oil and water needs to be prepared. In these cases, it may be more useful to study mixtures of oil (O), water (W), and emulsifier (E), as plots of ratios (Figure 9.4). The region of most suitable emulsion can be determined by studying varying mixtures.

9.2.6 Emulsion Stability and Analysis

The stability of emulsions is dependent on various parameters, including size of drops and interactions between drops. These different parameters are described in the following.

9.2.6.1 Emulsion Drop Size Analysis

Since the stability and other characteristics (such as viscosity and appearance) are known to be related to the drop size, these need to be measured. There are commercial instruments useful for such analyses. These different instruments are:

- *Coulter counter*—This is the most common type of apparatus where one simply counts the number of particles or drops passing through a well-defined hole. A signal is produced that corresponds to the size of the particle.
- *Light scattering*—Modern laser-light-scattering instruments are very advanced for particle size distribution analysis. The laser light is scattered by the small dispersed particles or drops. The latter is known to be dependent on the radius of particle. Emulsions are stable as long as the drops are separated from one another. Flocculation of an emulsion or dispersion takes place upon collision of the droplets, which is related to the Brownian motion, convective stirring, or gravitational forces. Actually, any emulsion can be separated into the oil and water phases by suitable centrifugation treatment.

9.2.6.2 Electrical Emulsion Stability

In those systems where the emulsifier carries a charge, it would impart specific characteristics to the emulsion. A double layer exists around the oil droplets in an O/W emulsion (Birdi, 2010a). If the emulsifier is negatively charged, then it will attract positive counterions while repelling negative charged ions in the water phase. The change in potential at the surface of oil droplets will be dependent on the concentration of ions in the surrounding water phase. The state of stability under these conditions can be qualitatively described as follows. As two oil droplets approach each other, the negative charge gives rise to a repulsive effect (see Chapter 7, Figure 7.4). The repulsion will take place within the electrical double layer (EDL) region. It can thus be seen that the magnitude of EDL distance will decrease if the concentration of ions in the water phase increases (Birdi, 2010b). This is due to the fact that the electrical double layer region decreases. However, in all such cases where two bodies come closer, two different kinds of forces exist that must be considered:

$$\text{Total force between two bodies} = \text{Repulsion forces} + \text{Attraction forces}$$

The nature of the total force thus determines whether:

- The two bodies will stay apart
- The two bodies will merge and form a conglomerate

This is a very simplified picture but a more detailed analysis will be given elsewhere. The attraction forces arise from van der Waals forces. The kinetic movement will finally determine whether the total force can maintain contact.

9.2.7 Orientation of Molecules at Oil–Water Interfaces

Currently, there is no method available by which one can directly determine the orientation of molecules of liquids at interfaces. Molecules are situated at interfaces (e.g., air–liquid, liquid–liquid, solid–liquid) under asymmetric forces. Recent studies have been carried out to obtain information about molecular orientation from surface tension studies of fluids (Birdi, 1997). It has been concluded that interfacial water molecules in the presence of charged amphiphiles are in a tetrahedral arrangement similar to the structure of ice. Extensive studies of alkanes near their freezing point had indicated that surface tension changes in abrupt steps. X-ray scattering of liquid surfaces indicated similar behavior (Wu et al., 1993). However, it was found that lower chain alkanes (C16) did not show this behavior. The crystallization of C16 at 18°C shows an abrupt change due to the contact angle change at the liquid–Pt plate interface (Birdi, 1997). It has been found that in comparison to C16–air interface one observes supercooling (to about 16.4°C). Each data point corresponded to 1 second, thus the data showed that crystallization is very abrupt. High-speed data (<<1 sec) acquisition is needed to determine the kinetics of transition. This kinetic data would thus add more information about molecular dynamics at interfaces and effect of additives to aqueous phase, such as proteins. The magnitude of IFT is 12.6 and 4 mN/m for bovine serum albumin (BSA) and casein, respectively (Birdi, 2002).

9.3 MICROEMULSIONS

As mentioned earlier, ordinary emulsions as prepared by mixing oil, water, and emulsifier are thermodynamically unstable (Birdi, 2009, 2010a). In other words, such an emulsion may be stable over a long period of time, but it will finally separate into two phases (oil phase and aqueous phase). All such emulsions can be separated into two phases, that is, the oil phase and water phase, by centrifugation. These emulsions are opaque which means that the dispersed phase (oil or water) is present in the form of large droplets (over micrometers and thus visible to the naked eye).

A *microemulsion* is defined as a thermodynamically stable and a clear isotropic mixture of water, oil, surfactant, and cosurfactant (in most systems it is short-chain alcohol). The cosurfactant is the fourth component that gives rise to the formation of very small aggregates or drops, which makes the microemulsion almost clear. Microemulsions are also therefore characterized as microstructured, thermodynamically stable mixtures of water, oil, surfactant, and additional components (such as cosurfactants). Intensive studies of microemulsions have shown that these are one of the following types:

- Microdroplets of oil in water or water in oil
- Bicontinuous structure

Emulsifier will be found in both these phases. On the other hand, in systems with four components (Figure 9.5) consisting of oil, water, detergent, and cosurfactant, a region exists where the clear phase is found. This is the phase region where microemulsions are found. Microemulsions are thermodynamically stable mixtures. The interfacial tension is almost zero. The size of drops is very small and this makes

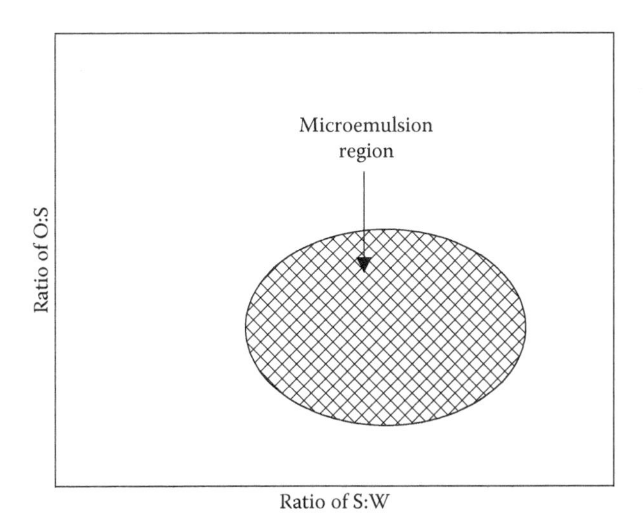

FIGURE 9.5 Four component system: oil (O)–water (W)–emulsifier (E)–cosurfactant (S) (ratio of O:S versus S:W).

microemulsions look clear. It has also been suggested that microemulsions may consist of bicontinuous structures. This sounds more plausible in these four-component microemulsion systems. It has been suggested that microemulsion may be compared to swollen micelles (that is, if one solubilizes oil in micelles). In such isotropic mixtures a short-range order exists between the droplets. It has been found from extensive experiments that not all mixtures of water, oil, surfactant, and cosurfactant give rise to a microemulsion. This has led to some studies that have tried to predict the molecular relationship.

Microemulsions have been formed using the following procedure:

- Oil–water mixture is added to a surfactant. To this emulsion one keeps adding a short-chain alcohol (with four to six carbon atoms) until a clear mixture (microemulsion) is obtained.

Thus, microemulsions will exhibit special properties, quite different than those exhibited by the ordinary emulsions. The microdrops may be considered as large micelles.

The literature related to microemulsions is extensive (Friberg et al., 2003; Birdi, 2009, 2010a). Accordingly, only a few typical examples will be delineated here in detail. A very typical microemulsion extensively investigated consists of a mixture of:

$$SDS + C_6H_6 + Water + Cosurfactant\ (C_5OH\ or\ C_6OH)$$

The phase region is determined by mixing various mixtures (approximately 20 samples) and allowing the system to reach equilibrium under controlled temperatures. From literature one finds the following recipe (Birdi, 1982): Mix 0.0032 mol (0.92 g) SDS (mol. Wt. of SDS ($C_{12}H_{25}SO_4Na$) = 288) with 0.08 mol (1.44 g) water and add 40 ml of C_6H_6. This mixture is mixed by vigorously stirring and one gets a creamy emulsion. While stirring this three-component mixture, a cosurfactant ($C_5H_{11}OH$ or $C_6H_{13}OH$) is added slowly until a clear system consisting of a microemulsion is obtained. The stability region is found to be a relation between surfactant–water and surfactant–alcohol. This shows that some kind of structure (at molecular level) is responsible. This shows that liquid crystal structure is indeed involved. The size of oil droplets is under a micrometer and therefore the mixture is clear (Birdi, 1982, 2009, 2010a) as seen by naked eye. These data clearly indicate that the microemulsion phase was formed at certain fixed surfactant:water and cosurfactant:oil ratios.

It is important to consider the different stages when proceeding to microemulsions from macroemulsions. As mentioned earlier, surfactant molecules orient with the hydrophobic group inside the oil phase, while the polar group orients toward the water phase. The orientation of surfactants at such interfaces cannot be measured by any direct method. Much useful information can be obtained from monolayer studies of air–water interfaces or oil–water interfaces.

At present, it is generally accepted that it is not easy to predict a microemulsion recipe. However, some suggestions have been offered:

- The HLB value of the surface needs be determined (for deciding the O/W or W/O type).
- The phase diagram of the water–oil–surfactant (and cosurfactant) needs to be determined.
- The effect of temperature is very crucial.
- The effect of added electrolytes is of additional importance.

In a recent report, the phase equilibriums of a microemulsion were reported. Phase behavior of a microemulsion formed with food grade surfactant sodium bis-(2-ethylhexyl) sulfosuccinate (AOT) was studied. Critical microemulsion concentration was deduced from the dependence of pressure on cloud points on the concentration of surfactant AOT at a constant temperature and water concentration. The results show that there are transition points on the cloud point curve in a very narrow range of concentration of surfactant AOT. The transition points were changed with the temperature and water concentration. These phenomena show that a lower temperature is suitable to forming microemulsion droplets and the microemulsion with high water concentration is likely to absorb more surfactants to the structure of the interface.

9.3.1 Some Typical Emulsion Recipes

Systems (oil and water) in which one phase is dispersed in another are found in many areas in everyday life, such as floor waxes, shaving creams, beverages, pesticide dispersions, skin creams, and pharmaceutical preparations. In the following some typical emulsion examples are given, which should give some data about these rather complex systems. In most emulsion products, the stability and shelf life is actually investigated by real-life tests.

9.3.1.1 Cleaning and Polishing Emulsions

In the application of emulsions for cleaning and polishing, a water phase and an oil phase (some selected halogenated solvent) are needed. The emulsions are used in the cleaning and polishing of surfaces of metals and other hard surfaces (e.g., glass).

For cleaning grease from metal surfaces, strong alkali media are needed (e.g., soda ash, borax or alkali phosphates or silicates). But these alone cannot effectively remove baked-on grease from metal surfaces. However, mixtures of soda and emulsifiers are effective.

9.3.1.2 Microemulsion Detergents

The application of microemulsions is rapidly growing. In the following some examples are given that indicate the methodology of applying microemulsions.

A light-duty microemulsion liquid detergent composition, useful for removing greasy soils from surfaces with both neat and diluted forms of the detergent composition, has been reported. It consists of the following components:

- 1% to 10% of a moderately water soluble complex of anionic and cationic surfactants, in which the complex (the anionic and cationic moieties) is in essentially equivalent or equimolar proportions

- An anionic detergent
- 1% to 5% cosurfactant
- 1% to 5% organic solvent
- 70% water

The recipe is based upon the following considerations. It is known that if an anionic (such as SDS) detergent is mixed with a cationic (such as CTAB), then a complex (1 mole:1 mole) is formed that is sparingly soluble in water. The reason being that positively and negatively charged moieties interact and produce a neutral complex, which is insoluble in water. This complex is oil soluble.

The complex component is one in which the anionic and cationic moieties include hydrophilic portions or substituents, in addition to the complex forming portions thereof. The anionic detergent is a mixture of higher paraffin sulfonate and higher alkyl polyoxyethylene sulfate. The cosurfactant is a polypropylene glycol ether, a poly-lower alkylene glycol lower alkyl ether or a poly-lower alkylene glycol lower alkanoyl ester. The organic solvent is a nonpolar oil, such as an isoparaffin, or an oil having polar properties, such as a lower fatty alkyl chain.

This liquid detergent has been reported to be an effective light-duty microemulsion liquid detergent composition, which is useful for the removal of greasy soils from substrates, both in neat form and when diluted with water.

9.3.2 Characteristics and Stability of Emulsions

The stability of any emulsion is dependent on the needs and the application area. In some cases, the emulsion needs to be stable for longer time than in other cases. As in the case of hair cream, the emulsion should destabilize as soon as it is applied to the hair. Otherwise, the hair would be white with emulsion droplets. On the other hand, any emulsion used in spraying on plants needs to be stable for longer time. Further, if an oil spill on an ocean needs to be cleaned, the emulsion formation needs to be destabilized.

There are different processes involved in the stability. One process is *the* creaming or flocculation of drops. This process is described in those cases where oil drops (in the case of oil–water) cling to one another and grow in large clusters. The drops do not merge. The density of most oils is lower than that of water. This leads to the fact that instability causes the oil drop clusters to rise to the surface (Figure 9.6).

This process can be reduced by increasing the viscosity of the water phase and thereby decreasing the rate of movement of the oil drops, and by decreasing interfacial tension and thus the size of the oil drops. The ionized surfactants will stabilize the O/W emulsions by imparting surface EDL.

9.4 LYOTROPIC LIQUID CRYSTALS

A solid melts when the temperature is increased more than its melting point. In the case of pure solids (almost 100%) this transition is very prompt. However, there are certain systems where the melting state can be very different than the ordinary solid-to-liquid transition.

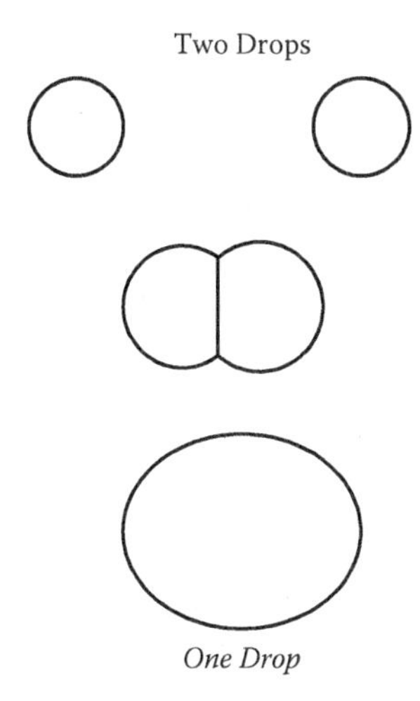

FIGURE 9.6 (Top) Two oil drops, (middle) flocculation step, (bottom) coalescence.

Liquid crystals (LCs) are phase structures that are intermediate between *liquid* and *crystal* phases (Friberg et al., 2003; Soltis et al., 2004). LCs have also been mentioned as mesophases (Greek *mesos* which means "middle"). LCs may be described as follows. If a pure substance, such as stearic acid, is heated it melts at a very specific temperature. Actually, the LC systems arise from the self-assembly characteristics as found in most of the amphiphile molecules. Heating a pure solid shows the following behavior:

Solid → Melting point → Liquid

However, if an LC substance is heated, it will show more than one melting point.

Solid → Melting point → Liquid1 → Melting point → Liquid2

Thus, LCs are substances that exhibit a phase of matter that has properties between those of a conventional liquid and those of a solid crystal. For instance, an LC may flow like a liquid but have the molecules in the liquid arranged or oriented in a crystal-like way. There are many different types of LC phases that can be distinguished based on their different optical properties (such as birefringence). When viewed under a microscope using a polarized light source, different liquid crystal phases will appear to have a distinct texture. Each "patch" in the texture corresponds to a domain where the LC molecules are oriented in a different direction. Within a domain, however, the molecules are well ordered. Liquid crystal materials may not always be in an LC phase (just as water is not always in the liquid phase, it may also be found in the solid or gas phase). Liquid crystals can be divided into the following types: thermotropic LCs and lyotropic LCs. *Thermotropic LCs* exhibit a phase transition into the LC phase as the temperature is changed, whereas *lyotropic LCs* exhibit phase transitions as a function of concentration of the mesogen in a solvent (typically water) as well as temperature.

9.4.1 Liquid Crystal Phases

The various LC phases (called mesophases) can be characterized by the type of molecular ordering that is present. One can distinguish positional order (whether molecules are arranged in any sort of ordered lattice) and orientational order (whether molecules are mostly pointing in the same direction), and moreover order can be either short range (only between molecules close to each other) or long range (extending to larger, sometimes macroscopic, dimensions). Most thermotropic LCs will have an isotropic phase at high temperatures. That is, heating will induce them into a conventional liquid phase characterized by random and isotropic molecular ordering (little to no long-range order), and fluid-like flow behavior. Under other conditions (for instance, lower temperatures), an LC might inhabit one or more phases with significant anisotropic orientational structure and long-range orientational order while still having an ability to flow. The ordering of liquid crystalline phases is extensive on the molecular scale. In fact, the self-assembly characteristic as possessed by lipids (amphiphiles) is the basic building feature in liquid crystals. This order extends up to the entire domain size, which may be on the order of micrometers but usually does not extend to the macroscopic scale as often occurs in classical crystalline solids. However, some techniques (such as the use of boundaries or an applied electric field) can be used to enforce a single-ordered domain in a macroscopic liquid crystal sample. The ordering in a liquid crystal might extend along only one dimension, with the material being essentially disordered in the other two directions.

Thermotropic phases are those that occur in a certain temperature range. If the temperature is raised too high, thermal motion will destroy the delicate cooperative ordering of the LC phase, pushing the material into a conventional isotropic liquid phase. At too low a temperature, most LC materials will form a conventional (though anisotropic) crystal. Many thermotropic LCs exhibit a variety of phases as the temperature is changed. For instance, a particular mesogen may exhibit various smectic and nematic (and finally isotropic) phases as the temperature is increased.

One of the most common LC phases is the *nematic phase*, where the molecules have no positional order, but they do not have long-range orientational order. Thus, the molecules flow and their center of mass positions are randomly distributed as in a liquid, but they all point in the same direction (within each domain). Most nematics are uniaxial: they have one axis that is longer and preferred, with the other two being equivalent (can be approximated as cylinders). Some liquid crystals are biaxial nematics, meaning that in addition to orienting their long axes, they also orient along a secondary axis. Liquid crystals are a phase of matter whose order is intermediate between that of a liquid and that of a crystal. The molecules are typically rod-shaped organic moieties about 25 Å (2.5 nanometers) in length and their ordering is a function of temperature. The nematic phase, for example, is characterized by the orientational order of the constituent molecules. The molecular orientation (and hence the material's optical properties) can be controlled with applied electric fields. Nematics are (still) the most commonly used phase in liquid crystal displays (LCDs), with many such devices using the twisted nematic geometry. The smectic phases, which are found at lower temperatures than the nematic, form well-defined layers that can slide over one another like soap. The smectics are thus positionally ordered along one direction. In the smectic A phase, the molecules are oriented along the layer normal, whereas in the smectic C phase they are tilted away from the layer normal. There is a very large number of different smectic phases, all characterized by different types and degrees of positional and orientational order.

9.4.2 Lyotropic Liquid Crystals

As compared to the cholesteric LC, the lyotropic LC consists of two or more components that exhibit liquid-crystalline properties (dependent on the concentration, temperature, and pressure). In the lyotropic phases, *solvent molecules* fill the space around the compounds (such as soaps) to provide fluidity to the system. In contrast to thermotropic liquid crystals, these lyotropics have another degree of freedom of concentration that enables them to induce a variety of different phases. A typical lyotropic liquid crystal is a surfactant, water, and a long-chain alcohol.

A compound that has two immiscible hydrophilic and hydrophobic parts within the same molecule is called an amphiphilic molecule (as mentioned earlier). Many amphiphilic molecules show lyotropic liquid–crystalline phase sequences depending on the volume balances between the hydrophilic part and hydrophobic part. These structures are formed through the microphase segregation of two incompatible components on a nanometer scale. Hand soap is an everyday example of a lyotropic liquid crystal (80% soap + 20% water). The content of water or other solvent molecules changes the self-assembled structures. At very low amphiphile concentrations, the molecules will be dispersed randomly without any ordering. At a slightly higher (but still low) concentration, amphiphilic molecules will spontaneously assemble into micelles or vesicles. This is done so as to "hide" the hydrophobic tail of the amphiphile inside the micelle core, exposing a hydrophilic (water-soluble) surface to aqueous solution. These spherical objects do not order themselves in solution, however. At higher concentration, the assemblies will become ordered. A typical phase is a hexagonal columnar phase, where the amphiphiles form long cylinders (again with a hydrophilic surface) that arrange themselves into a roughly hexagonal lattice. This is called the middle soap phase. At still higher concentrations, a lamellar phase (neat soap phase) may form, wherein extended sheets of amphiphiles are separated by thin layers of water. For some systems, a cubic (also called viscous isotropic) phase may exist between the hexagonal and lamellar phases, wherein spheres are formed that create a dense cubic lattice. These spheres may also be connected to one another, forming a bicontinuous cubic phase. The aggregates created by amphiphiles are usually spherical (as in the case of micelles), but may also be disc-like (bicelles), rod-like, or biaxial (all three micelle axes are distinct) (Zana, 2008). These anisotropic self-assembled nanostructures can then order themselves in much the same way as liquid crystals do, forming large-scale versions of all the thermotropic phases (such as a nematic phase of rod-shaped micelles).

These structures are described extensively in the current literature (Birdi, 2002; Friberg et al., 2003; Somasundaran, 2006; Fanum, 2008). Even within the same phases, their self-assembled structures are tunable by the concentration. For example, in lamellar phases, the layer distances increase with the solvent volume. Lamellar structures are found in systems such as common hand soap. The latter consists of about 0% soap and 20% water. The layers of soap molecules are separated by a region of water (including salts) as a kind of sandwich (as found for x-ray diffraction analyses). Since lyotropic liquid crystals rely on a subtle balance of intermolecular interactions, it is more difficult to analyze their structures and properties than those of thermotropic liquid crystals. Similar phases and characteristics can be observed in immiscible diblock copolymers. As mentioned earlier, surfactant aggregate form micelles that may vary in size (i.e., number of monomers per micelle) from a few to over a thousand monomers. However, surfactants can form in addition to simple micellar aggregates (i.e., spherical or ellipsoidal) many other structures when mixed with other substances. The curved micelle aggregates are known to change to planar interfaces when additives, so-called cosurfactants, are added. A procedure reported is a recipe consisting of:

Surfactant (cetylpyridinium bromide, CPBr) + Hexanol + Salt + Water

The addition of salts to micelles results in large micelles that turn into cylindrical shapes. However, the addition of cosurfactant gives rise to the liquid crystal phase. As a consequence, these micellar systems with added cosurfactant are found to undergo several macroscopic phase transitions in dilute solutions. These transitions are as follows:

Surfactant::Micelles (Spherical— — Cylindrical)
Surfactants + Cosurfactant::— — Liquid crystal

Further, the extensive change of ionic charges (if present) are very prominent. The simple picture that bilayers in aqueous medium are principally stabilized by the competition between *hydration forces* (van der Waals) and *electrostatic interactions* seems to be the most plausible basis. The hydration forces arise from the hydrogen bond formation between the polar groups of the surfactant and the water molecule. Van der Waals forces are attraction forces between all molecules (these are short range).

LC phases can also be based on low-melting inorganic phases like $ZnCl_2$ that have a structure formed of linked tetrahedral and easily formed glasses. The addition of long-chain soap-like molecules leads to a series of new phases that show a variety of liquid crystalline behavior both as a function of the inorganic–organic composition ratio and of temperature. This class of materials has been named *metallotropic.*

It would be no exaggeration to state that the whole biological world is basically based on LC structures. This is easily noticed when considering some simple molecules as found in nature, especially amphiphiles, which are found to exhibit characteristics that are very important for the biological structure and function. The effect of temperature on biological living systems needs a few remarks. The human body temperature is made up of molecules and cell structures such as to function at optimum at 37°C. If the temperature changes a few degrees (plus or minus) from 37°C then the effect is considerable but not lethal. This is due to the fact that the system shows LC behavior and keeps functioning (without an abrupt change). In other words the biological reactions can go on functioning (though with some restrictions) even if the temperature is 36°C or 38°C. Nature uses this structure in order to be able to sustain life over a range of temperatures.

Lyotropic liquid-crystalline nanostructures are abundant in living systems. Accordingly, lyotropic LC have been of much interest in such fields as biomimetic chemistry. In fact, biological membranes and cell membranes are a form of LC. Their constituent rod-like molecules (e.g., phospholipids) are organized perpendicularly to the membrane surface, yet the membrane is fluid and elastic. The constituent molecules can flow in-plane quite easily but tend not to leave the membrane, and can flip from one side of the membrane to the other with some difficulty. These LC membrane phases can also host important proteins such as receptors freely "floating" inside or partly outside the membrane.

Many other biological structures exhibit LC behavior. For instance, the concentrated protein solution that is extruded by a spider to generate silk is actually an LC phase. The precise ordering of molecules in silk is critical to its renowned strength. DNA and many polypeptides can also form LC phases. Since biological mesogens are usually chiral; chirality often plays a role in these phases.

9.4.3 Theory of Liquid Crystal Formation

Microscopic theoretical treatment of fluid phases can become quite involved, owing to the high material density, which means that strong interactions, hard-core repulsions, and many-body correlations cannot be ignored. In the case of LCs, anisotropy in all of these interactions further complicate analysis. There are a number of fairly simple theories, however, that can at least predict the general behavior of the phase transitions in LC systems. Additionally, the description of LCs involves an analysis of order. In particular, a sharp drop of the order parameter to zero is observed when one undergoes a phase transition from an LC phase into the isotropic phase. The order parameter can be measured experimentally in a number of ways. For instance, diamagnetism, birefringence, Raman scattering, nuclear magnetic resonance (NMR) spectroscopy, and electron paramagnetic resonance (EPR) spectroscopy can also be used to determine the order parameter. A very simple model that predicts lyotropic phase transitions is the hard-rod model proposed (Friberg et al., 2003). This theory considers the volume excluded from the center of mass of one idealized cylinder as it approaches another. Specifically, if the cylinders are oriented parallel to one another, there is very little volume that is excluded from the center of mass of the approaching cylinder (it can come quite close to the other cylinder). If, however, the cylinders are at some angle to one another, then there is a large volume surrounding the cylinder where the approaching cylinder's center of mass cannot enter (due to the hard-rod repulsion between the two idealized objects). Thus, this angular arrangement sees a decrease in the net positional entropy of the approaching cylinder (there are fewer states available to it). It may be asserted that the fundamental reason arises from the fact that, while parallel arrangements of anisotropic objects leads to a decrease in orientational entropy, there is an increase in positional entropy. Thus, in some case greater positional order will be entropically favorable. This theory thus predicts that a solution of rod-shaped objects will undergo a phase transition, at sufficient concentration, into a nematic phase.

According to one theoretical model one assumes that the liquid crystal material is a continuum. It has been suggested that three types of distortions can take place in these structures:

1. Twists of the material, where neighboring molecules are forced to be angled with respect to one another
2. Nonlinear (or splay) of the material, where bending occurs perpendicular to the director
3. Bend of the material, where the distortion is parallel to the director and mesogen axis

---

All three of these types of distortions incur an energy penalty. They are defects that often occur near domain walls or boundaries of the enclosing container. The response of the material can then be decomposed into terms based on the elastic constants corresponding to the three types of distortions. It has been mentioned in the literature that chiral mesogens usually give rise to chiral mesophases. For molecular mesogens, this means that the molecule must possess some form of asymmetry, usually a stereogenic center. An additional requirement is that the system not be racemic; a mixture of right- and left-handed versions of the mesogen will cancel the chiral effect. Due to the cooperative nature of LC ordering, however, a small amount of chiral dopant in an otherwise achiral mesophase is often enough to select out one domain handedness, making the system overall chiral. Chiral phases usually have a helical twisting of the mesogens. If the pitch of this twist is on the order of the wavelength of visible light, then interesting optical interference effects can be observed. The chiral twisting that occurs in chiral LC phases also makes the system respond differently to right- and left-handed circularly polarized light. These materials can thus be used as polarization filters. It is possible for chiral mesogens to produce essentially achiral mesophases. For instance, in certain ranges of concentration and molecular weight, DNA will form an achiral line hexatic phase. A curious recent observation is of the formation of chiral mesophases from achiral mesogens. Specifically, bent-core molecules (sometimes called banana liquid crystals) have been shown to form LC phases that are chiral. In any particular sample, various domains will have opposite handedness, but within any given domain, strong chiral ordering will be present.

9.4.4 Industrial Applications of Liquid Crystals

Liquid crystals find a wide use in LC displays, which rely on the optical properties of certain liquid-crystalline molecules in the presence or absence of an electric field or temperature gradient. In a typical device, an LC layer sits between two polarizers that are crossed (oriented at 90° to one another). The LC is chosen so that its relaxed phase is a twisted one. This twisted phase reorients light that has passed through the first polarizer, allowing it to be transmitted through the second polarizer and reflected back to the observer. The device thus appears clear. When an electric field is applied to the LC layer, all the mesogens align (and are no longer twisting). In this aligned state, the mesogens do not reorient light, so the light polarized at the first polarizer is absorbed at the second polarizer, and the entire device appears dark. In this way, the electric field can be used to make a pixel switch between clear or dark on command. Color LCD systems use the same technique, with color filters used to generate red, green, and blue pixels. Similar principles can be used to make other LC-based optical devices.

Thermotropic chiral LCs whose pitch varies strongly with temperature can be used as crude thermometers, since the color of the material will change as the pitch is changed. LC color transitions are used on many aquarium and pool thermometers. Other LC materials change color when stretched or stressed. Thus, LC sheets are often used in industry to look for hot spots, map heat flow, measure stress distribution patterns, and so on. LC in fluid form is used to detect electrically generated hot spots for failure analysis in the semiconductor industry. Liquid crystal memory units with extensive capacity were used in space shuttle navigation equipment. It is also worth noting that many common solutions are in fact LCs. Soap, for instance, is an LC, and forms a variety of LC phases depending on its concentration in water (Friberg et al., 2003; Somasundaran, 2006).

9.5 APPLICATIONS OF EMULSIONS

The area of emulsion applications is very large and cannot be described in the space available here. However, some important areas will be highlighted to indicate the basic criteria for emulsion technology (Friberg et al., 2003; Gitis and Sivamani, 2004).

9.5.1 Cosmetics and Personal Care Industry

Human skin is the outer layer of the body that protects the latter from any exposure to the surroundings (wind or rain or mechanical strain). The natural substances that compose the skin are very elaborate and complex. Further, the composition of skin changes with age, and is different for every person. The same is true for the animal world. The skin is maintained in communication with the underlying part of the body through pores. A continuous transport of various substances through these pores (such as water, salts, and fats) exists. Thus, an equilibrium exists between these outer layers and the inside body. However, in some cases this equilibrium may not be optimum, which leads to skin irritation (red skin or rashes or itching). This requires treatment of the skin by using cosmetics that are suitable to cure the skin. It is known that at present the personal care industry is one of the largest worldwide (Birdi, 2010a). The skin barrier properties and effect of hand hygiene practices are known to be of important concern. The average adult has a skin area of about 1.75 m^2. The superficial part of the skin, the *epidermis*, has five layers. The *stratum corneum*, the outermost layer, is composed of flattened dead cells (corneocytes or squames) attached to each other to form a tough, horny layer of keratin mixed with several lipids, which help maintain the hydration, pliability, and barrier effectiveness of the skin. This part of skin has been compared to a wall of bricks (corneocytes) and mortar (lipids), and serves as the primary protective barrier. Approximately 15 layers make up the stratum corneum, which is completely replaced every 2 weeks; a new layer is formed approximately daily. From healthy skin, approximately 10^7 particles are disseminated into the air each day, and 10% of these skin squames contain viable bacteria. This is a source of major dirt inside the house and contributes to many interactions. Skin is the organ that covers and protects the human body. This is also the same as for all other living species. The other functions are regulation of the body temperature (about 37°C) and also the penetration control (for example, sunlight, liquids, and solid materials). Thus, the role of skin is important for all living beings. For example, for large domestic animals like buffalo, water intake is critical for sustaining life. Human skin is composed of two layers of different tissue:

1. Underlying layer—Subcutaneous tissue.
2. Dermis or true skin—In this region the sensory nerves, blood vessels and sweat glands are located. In this region a fatty substance sebum is created, which coats, lubricates, and lets other molecules to pass through. This also affects water loss.

If for some reason, the outer part of skin is damaged, then special creams and ointments for such repair purpose are found commercially. These are described in the following.

9.5.1.1 Fundamentals of Skin Creams and Recipes

It is thus clear from the preceding discussion that any reaction between cosmetic preparation and the components of the skin need to be considered. Some typical properties and recipes for skin cream cover a very large area of information. Skin creams are known to be composed of a variety of ingredients, which are based on the end use (hands, feet, face, hair, etc.). Some real specialty products are applied for dry or repair skin effects. In the following some recipes are given. The aim of the different cosmetics is to repair and restore the original balance in the skin structure. Since the number of personal care emulsion creams is so large, only some typical examples are given here. A variety of emulsion skin care products that claim to exhibit properties for this purpose are found commercially. In some cosmetics it has been found that proteins can have both moisturizing and tightening agents. Mankind has realized that hygiene of the skin is related to a clean state.

A typical ingredient list for a facial cream (dry skin):

- Water
- Safflower seed oil
- Coconut oil
- Glycerine polysorbate 60
- Evening primrose oil

- Sorbitan stearate
- Cetyl ester
- Ethylenediaminetetraacetic acid (EDTA)

Additionally, many procedures have been suggested to the assessment of dry skin. Evaporation of emulsion water, electrical capacitance, and skin surface (emulsion) lipids are methods used for the assessment of the efficacy of skin-care products. Hygiene of the skin is one of the most important aspects of daily life. One may also ask the question how clean is clean. Environmental sanitation and public health services, despite room for improvement, are generally good. In addition, choices of hygienic skin care products have never been more numerous, and the public has increasing access to health- and product-related information. It has been reported that there is evidence for the relationship between skin hygiene and infection, the effects of washing on skin integrity, and recommendations for skin-care practices for the public and health care professionals. Such agents, unlike plain soaps, reduce microbial counts on the skin. Whole-body washing with chlorhexidine-containing detergent has been shown to reduce infections among neonates, but concerns about absorption and safety preclude this as a routine practice. These factors have led to suggestions that antimicrobial products should be more universally used, and a myriad of antimicrobial soaps and skin-care products have become commercially available. Although antimicrobial drug-containing products are superior to plain soaps for reducing both transient pathogens and colonizing flora, widespread use of these agents has raised concerns about the emergence of bacterial strains resistant to antiseptic ingredients, such as triclosan. Some evidence indicates that long-term use of topical antimicrobial agents may alter skin flora. Water content, humidity, pH, intracellular lipids, and rates of shedding help retain the protective barrier properties of the skin. When the barrier is compromised (e.g., by hand hygiene practices such as scrubbing), skin dryness, irritation, cracking, and other problems may result.

The demand for hand hygiene among health care professionals has never been greater than in modern times. A mild emulsion cleansing rather than hand washing with liquid soap was associated with a substantial improvement in the skin of nurses' hands.

Moisturizing is beneficial for skin health and reducing microbial dispersion from skin, regardless of whether the product used contains an antibacterial ingredient. Because of differences in the content and formulations of lotions and creams, products vary greatly in their effectiveness. Lotions used with products containing chlorhexidine gluconate must be carefully selected to avoid neutralization by anionic surfactants. Further, various LC applications in skin care cosmetics have been also reported.

Intercellular lipids of the stratum corneum contribute threefold to the maintenance of a healthy skin, by hydration, cell adhesion, and reduction of transepidermal water loss. All of these functions can be attributed to the self-assembly property of the amphiphilic molecules of the stratum corneum lipids.

In recent years, solid lipid nanoparticles (SLN) have been introduced as a novel carrier system for drugs and cosmetics. It has been found that SLN possess characteristics of physical UV blockers on their own, thus offering the possibility of developing a more effective sunscreen system with reduced side effects. Aqueous SLN dispersions were produced and incorporated into gels, followed by particle size examination, stability testing upon storage, and thermoanalytical examination. Investigation of the UV-blocking capacity using different in vitro techniques revealed that the SLN dispersions produced in this study are at least twice as effective as their reference emulsions (conventional emulsions with identical lipid content). Furthermore, film formation of SLN on the skin and occlusivity were examined. The influence of a cream containing 20% glycerine and its vehicle on skin barrier properties has been investigated. Recent studies have shown that polymers offer several advantages and can be used in skin-care products. The phase diagrams were determined of lactic and isohexanoic hydroxy acids as well as salicylic acid with water, a nonionic surfactant, and a paraffinic oil to outline the influence of the hydroxy acids on the structure in a model for a skin lotion. The results showed the influence of the acid to be similar to that of the oil, but that the difference in chain length between the two alpha acids had only insignificant influence. The results are discussed from two aspects: the structures involved in the lotion as applied and the action of the lotion residue on the skin after the evaporation of the water. In pharmaceutics, an application-triggered drug release from an O/W-emulsion recipe has been reported in the literature.

9.5.2 Paint Industry and Colloid Aspects

Everyone appreciates the role of paint, on cars or houses, or wood furniture, and so on. The function of paint is manyfold:

- Appearance
- Color
- Protection (corrosion, in case of iron; wood)

Paint is recognized as a pigmented mixture (in liquid or paste form) that protects and enhances the appearance of materials (Birdi, 2009, 2010a). In the early days (almost 100 years ago), cars were painted with a nitrocellulose lacquer. Especially, in some areas (such as cars) the industrial development has been very wide ranging.

The paint industry is extensive. The various paint properties of significance are:

- Paint formulation
- Paint pigments
- Paint rheology and stability
- Paint stripper
- Paint pretreatment
- Polyester paint
- Polyurethane coating
- Metallizing
- Marine paint and corrosion
- Paint removal

Paints consist of essentially the following components:

- Pigment-metallic
- Binder
- Dispersants (emulsifiers, etc.)
- Solvent
- Additives

These mixtures are complex systems and much of the industrial development is safeguarded by the patents.

9.6 EMULSION STABILITY AND STRUCTURE

The degree of stability of any emulsion is related to the rate of coagulation of two drops (O/W, oil drops; W/O, water drops).

$$\text{Oil drop} + \text{Oil drop} \rightarrow \text{Time} \rightarrow \text{One oil drop}$$

The length of time is the degree of emulsion stability. This process means that two oil drops in an O/W emulsion come close together and if the repulsion forces are smaller than the attraction forces, only then the two particles meet and fuse into one larger drop. In the case of charged drops, there will be an EDL around these drops. A negatively charged oil drop (charge arising from the emulsifier) will strongly attract positively charged ions in the surrounding bulk aqueous phase. At a close distance from the surface of a drop the distribution of charges will change a lot. Whereas at a very large distance there will be electrical neutrality as there will be even number of positive and negative charges.

The electrostatic repulsion exists between the two negatively charged drops, which would exhibit strong repulsion even at large distances (many times the size of the particle). The shape of the EDL curve will be dependent on the negative and positive charge distribution. It is easily seen that if the concentration of counterions increases, then the magnitude of EDL will decrease and this will decrease the maximum of the total potential curve. The stability of emulsions can thus be increased by decreasing the counterion concentration. Another important emulsion stabilization is achieved by using polymers. The large polymer molecules adsorbed on solid particles will exhibit repulsion at the surface of particles. The charged polymers will thus also give additional charge–charge repulsion. Polymers are used in many pharmaceuticals, cosmetics, and other systems (e.g., milk). Obviously, the choice of a suitable polymer is specific for each system.

9.6.1 Diverse Emulsion Technology

As is already obvious, the emulsion technology is a state-of-art system and mostly protected by patents (Friberg et al., 2003; Sjoblom et al., 2008). In the following, some examples are given for the sake of information that can be of essential use.

9.6.1.1 Nanoemulsion Technology

The low-energy emulsification method has been used to prepare O/W *nanoemulsions* in the water–potassium oleate–oleic acid–C12E10–hexadecane system. This method has not been used practically in ionic systems until now. The resulting droplet sizes, much smaller than those obtained with the high-energy emulsification methods, depend on the composition and preparation variables (addition and mixing rate). Phase diagrams, rheology measurements, and experimental designs applied to nanoemulsion droplet sizes obtained were combined to study the formation of these nanoemulsions. To obtain nanosize droplets, it was necessary to cross a direct cubic liquid crystal phase along the emulsification path, and it is also crucial to remain in this phase long enough to incorporate all of the oil into the liquid crystal. When nanoemulsion forms, the oil is already intimately mixed with all of the components, and it only has to be redistributed. Results show that the smaller droplet sizes are obtained when the liquid crystal zone is wide and extends to high water content, because in this case, during the emulsification process, the system remains long enough in the liquid crystal phase to allow the incorporation of all of the oil. Around the optimal formulation variables, the LC zone crossed during emulsification is wide enough to incorporate all of the oil whatever mixing or stirring rate is used, and then the resulting droplet size is independent of preparation variables. However, when the composition is far from this optimum, the LC zone becomes narrower and the mixing of components controls the nanoemulsion formation.

9.6.1.2 Microemulsion Technology for Oil Reservoirs

A new microemulsion additive has been developed that is effective in remediating damaged wells and is highly effective in fluid recovery and relative permeability enhancement when applied in drilling and stimulation treatments at dilute concentrations. The microemulsion is a unique blend of a biodegradable solvent, surfactant, cosolvent, and water. The nanometer-sized structures are modeled with structures that when dispersed in the base, treating the fluid of water or oil permit a greater ease of entry into a damaged area of the reservoir or fracture system. The structures maximize surface energy interaction by expanding to 12 times their individual surface areas to allow maximum contact efficiency at low concentrations (0.1%–0.5%). Higher loadings on the order of 2% can be applied in the removal of water blocks and polymer damage. Laboratory data have shown that microemulsion speeds the cleanup of injected fluids in tight gas cores. Further tests show that the microemulsion additive results in lower pressures to displace fluids from propped fractures resulting in lower damage and higher production rates. This reduced pressure is also evident in pumping operations where friction is lowered by 10% to 15% when the microemulsion is added to fracturing fluids. Field examples are shown for remediation and fracture treating of coals, shales, and sandstone reservoirs, where productivity is increased by 20% to 50% depending on the treatment parameters. Drilling examples have been shown in horizontal drilling where wells clean up without the aid of workover rigs where offsets typically require weeks of workover.

Terpene-based microemulsion cleaning composition has been reported in some industrial applications. Oil-in-water microemulsion cleaning compositions comprising four principal components were described based upon four components. These were:

1. Terpene solvent, (e.g., d-limonene)
2. An aliphatic glycol monoether cosolvent (e.g., dipropylene glycol monomethyl ether)
3. A mixture of nonionic surfactants consisting of a capped alkylphenol ethoxylate or an ethoxylated higher aliphatic alcohol
4. A fatty acid alkanolamide and water

The cleaning composition may be used in a concentrated form or in a diluted form. The composition may be used for cleaning soil from among others glass and metal parts. This microemulsion shows that by combining water and oil one can clean metal surfaces effectively. This arises from the fact that both oil and water soluble dirt is removed by the microemulsion.

10 Essential Surface and Colloid Chemistry in Science and Industry

10.1 INTRODUCTION

As will be already obvious to the reader, the scope of applications of surface and colloid chemistry is extensive. These are systems where surfaces of liquids or solids are involved. There are systems where the particle size plays an important role. These characteristics thus require the need for knowledge of the physical forces involved. In this chapter some important systems, which are complex as compared to more fundamental examples described earlier, will be explained. This is useful, since scientific developments change as more theoretical knowledge is obtained under different subject. Another important point to mention is that in surface and colloid chemistry, the essential subject is the size, which ranges from more than a micrometer (10^{-6} m) to a nanometer (10^{-9} m). Further, with modern instruments, one can study these systems in greater detail, which becomes better every decade that goes by.

Essential topics of surface and colloid chemistry include:

- Curved surfaces
- Capillary pressure
- Particle size and surface area of solids
- Surface tension and interfacial tension of liquids
- Self-assembly structures of lipids
- Emulsions and interfacial tension
- Oil and gas recovery and capillary forces
- Solid surface tension
- Washing and cleaning technology
- Solid surfaces and catalytic processes

The examples discussed in this chapter are designed to indicate how surface and colloid chemistry have helped in the development of new applications in the recent years.

It also becomes evident that there will be many new areas where the application of surface and chemistry principles will be the determining factor, as new research reveals ever-increasing details. Some of these examples are given in this chapter as an illustration. For example, the capillary forces, as mentioned in Chapter 1, become extensively involved when considering the movement of water in a sponge. Sponge consists of many interconnected capillaries. Further, an oil reservoir can be considered as a simplified model of a sponge. If the reservoir is finely pored, then oil recovery is very poor (less than 30%), whereas if the pores are of large diameter then recovery will be very high (over 60%).

Additionally, there are also examples where surface and colloid chemical principles are implemented in complex systems. For example, the surface chemistry of bone structure is becoming an important area of research (Mousny et al., 2008). This area of applications is expanding as more fundamental understanding of the surface and colloid chemistry becomes evident.

In the recent decade, for example, the application of *nanoscience* and *nanotechnology* has developed since the molecular studies of surfaces at nanoscale have reached a very high level. Nanotechnology is nowadays a popular topic in materials science and other new areas are also emerging. It has deserved this popularity because of its interdisciplinary character, involving expertise in physics, chemistry, biology, and medicine. Many materials properties are studied in detail. The common factor in nanotechnology is the lateral dimension, being in the nanometer (10^{-9} m, that is 1 billionth of a meter or 1/1000th of the thickness of a paper sheet!) range, of the structures studied. Atomic or molecular distances, sizes of structures in semiconductor devices, and grain sizes in nanopowders are just a few examples out of many that can be considered as nanosized dimensions. In nanotechnology, knowledge of the structure and composition of the materials studied is a key requirement for understanding materials properties at a small distance (Birdi, 2003).

10.2 FOOD EMULSIONS (MILK INDUSTRY)

Man has learned much from those natural systems where surface and colloids have been found. One of the most important stages in life is the feeding of newborns. In this context, *milk* is one of the most amazing examples of emulsions found in nature, especially, when one considers the role of milk in the growth and nutrition of newborns. Milk consists of fats, nonfats (such as lactose), and inorganic salts. The milk emulsion has to be stable to a certain extent as well as it has to provide nutrition based on the many different molecules present in the emulsion (i.e., water, fats, proteins, inorganic salts, and organic molecules). Nature uses emulsion since it needs to provide fats and proteins as emulsion to newborns for nutrition and other needs, as well as to provide water and inorganic salts (Kristensen, 1997; Friberg et al., 2003). Because some children may need extra nutrition, synthetic milk is provided. Another example is ice cream, where the colloidal dispersion of ice particles are dispersed together with entrapped tiny air bubbles in the emulsion consisting of fats, sugar, and thickening agents (polysaccharides).

Composition of Milk and Ice Cream

	Ice Cream	Milk
Fats	13	4
Nonfats	10	8
Sugar	16	—
Emulsifier	0.4	—
Stabilizer	1	—
Inorganic salts	1	1

The structure of ice cream has been studied in detail by using electron microscopy. Trapped air bubbles are separated by only few micrometer-thick layers of the continuous phase. Ice cream consists of about 40% air frozen foam. The continuous phase in foam consists of sugars, proteins, and emulsifiers. Typical ice cream contains:

Fats, 13%
Milk, 10%
Sugar, 16%
Emulsifier, 0.4%
Salts, 1%

Ice cream emulsion has some very characteristic degrees of stability. The air bubbles should remain dispersed, but as soon as it melts in the mouth (due to high temperatures), the emulsion should break. This leads to the sensation of taste, which is very essential. The sensation (and recognition) of taste on the surface of tongue is known to be related to the molecules' shape and physicochemical properties. As soon as these molecules are separated from the emulsion the taste sensation is recorded in the brain. Therefore, the various components must stay in same phase after the break up of the emulsion. Emulsifiers that are generally used have low hydrophilic–lipophilic balance (HLB) values for water-in-oil. Furthermore, in food emulsions there are a variety of products where interfaces play an important role. One such example is churning, where one whips air bubbles in milk or cream. This process is the same as in the case of whipped cream. In the process, a large amount of air is incorporated into the liquid, in bubble form, and the fat globules (being surface active) collect in the bubble walls. This shows that surface active molecules, such as fats, are collected at the air–water interface during churning or whipping. But where whipping cream is kept cold, and the agitation stopped when a stable, airy foam is produced, churned cream is warmed to the point that the globules soften and to some degree liquefy. The ideal temperature range is said to be 12°C to 18°C. Persistent agitation knocks the softened globules into each other enough to break through the protective membrane, and liquid fat cements the exposed droplets together. The foam structure is broken both by the free fat and the released membrane materials, which include emulsifiers like lecithin. These materials disrupt thin water layers and so burst bubble walls, and once enough of them have been freed in the process of whipping or churning cream, the foam will never be stable again. As churning continues, then, the foam gradually subsides, and the butter granules are worked together into larger masses. Finally, the cream becomes thick and separates into butter grains and buttermilk. The size of the fat globules varies from 0.1 to 10 μm in diameter. The fat globule membrane is comprised of surface active materials: phospholipids and lipoproteins. Fat globules typically aggregate in three ways:

1. Flocculation
2. Coalescence
3. Partial coalescence

Actually, during the process of churning it can be concluded that the process of butter making can be described as an inversion of the original cream emulsion. The system of fat droplets dispersed in water is converted into a continuous phase of fat that contains water droplets. The final product is about 80% milk fat, 18% water, and 2% milk solid, mainly proteins and salts carried in the water.

10.2.1 Milk: Composition and Emulsion Chemistry

One of the most important natural systems of interest in surface chemistry is that of milk. The importance of milk in mammals is very well known. The role of milk in nature is to nourish and provide immunological protection for the mammalian young. Milk from humans, goats, buffalos, sheep, and yaks, has been a food source for humans since prehistoric times. It is not surprising, therefore, that the nutritional value of milk is high. Milk is also a very *complex food* where many thousands of different molecular species are found.

For example, an approximate composition of milk has been reported as 87% water; 4% milk fat; 9% solids nonfat; 3% protein; 5% lactose; and diverse small amounts of substances (minerals, enzymes, gases, and vitamins).

It is remarkable that milk is such a complex fluid yet the most significant natural nourishing substance (Kristensen, 1997). Milk may be described as follows:

- An oil-in-water emulsion with the fat globules dispersed in the continuous serum phase.
- A colloid suspension of casein micelles, globular proteins, and lipoprotein particles in a solution of lactose, soluble proteins, minerals, vitamins, and other components.

In image analyses of milk under a microscope, at low magnification (5×) a uniform but turbid liquid is observed. At 500× magnification, spherical droplets of fat, known as *fat globules*, can be seen. At even higher magnification (50,000×), the casein micelles can be observed. The main structural components of milk, fat globules and casein micelles, will be examined in more detail later. The main milk lipids are a class called triglycerides, which are comprised of a glycerol backbone binding up to three different fatty acids. The fatty acids are composed of a hydrocarbon chain and a carboxyl group. The major fatty acids found in milk are:

Long Chain	
$C_{14}H_{29}COOH$	Myristic 11%
$C_{16}H_{33}COOH$	Palmitic 26%
$C_{18}H_{37}COOH$	Stearic 10%
$C_{18}(1)H_{37}COOH$	Oleic 20%
Short Chain (11%)	
C_4H_9COOH	Butyric*
$C_6H_{13}COOH$	Caproic
$C_8H_{17}COOH$	Caprylic
$C_{10}H_{21}COOH$	Capric

* Butyric fatty acid is specific for milk fat of ruminant animals and is responsible for the rancid flavor when it is cleaved from glycerol by lipase action.

Saturated fatty acids (no double bonds), such as myristic, palmitic, and stearic make up two-thirds of milk fatty acids.

Oleic acid is the most abundant unsaturated fatty acid in milk with one double bond. Triglycerides account for 98% of milk fat. The small amounts of monoglycerides, diglycerides, and free fatty acids in fresh milk may be a product of early lipolysis or simply incomplete synthesis. Other classes of lipids include phospholipids (0.8%), which are mainly associated with the fat globule membrane, and cholesterol (0.3%), which is mostly located in the fat globule core.

At room temperature, the lipids are solid, and therefore are correctly referred to as "fat" as opposed to "oil," which is a liquid at room temperature. The final melting point of milk fat is 37°C because higher melting triglycerides dissolve in the liquid fat. This temperature is significant because 37°C is the body temperature of the cow, and the milk would need to be a liquid at this temperature.

10.2.1.1 Milk-Fat Structure: Fat Globules

More than 95% of the total milk lipid is in the form of a globule ranging in size from 0.1 to 15 μm in diameter. These liquid fat droplets are covered by a thin membrane, 8 to 10 nm in thickness, whose properties are completely different from both milk fat and plasma. The native fat globule membrane (FGM) is comprised of apical plasma membrane of the secretory cell, which continually envelops the lipid droplets as they pass into the lumen (Kristensen, 1997). The major components of the native FGM, therefore, are protein and phospholipids. The phospholipids are involved in the oxidation of milk. There may be some rearrangement of the membrane after release into the lumen as amphiphilic substances from the plasma adsorbed onto the fat globule, and parts of the membrane dissolve into either the globule core or the serum. The FGM decreases the lipid–serum interface to very low values, 1 to 2.5 mN/m, preventing the globules from immediate flocculation and coalescence, as well as protecting them from enzymatic action. It is well known that if raw milk or cream is left to stand, it will separate. IgM, an immunoglobulin in milk, forms a complex with lipoproteins. This complex, known as cryoglobulin, precipitates onto the fat globules and causes flocculation. This is known as cold agglutination. As fat globules cluster, the speed of rising increases and sweeps up the smaller globules with them. The cream layer forms very rapidly, within 20 to 30 minutes, in cold milk. Homogenization of milk prevents this creaming by decreasing the diameter and size distribution of the fat globules, causing the speed of rise to be similar for the majority of globules. As well, homogenization causes the formation of a recombined membrane, which is much similar in density to the continuous phase. Recombined membranes are very different than native FGM. Processing steps such as homogenization, decreases the average diameter of fat globule and significantly increases the surface area. Some of the native FGM will remain adsorbed, but there is no longer enough of it to cover all of the newly created surface area. Immediately after disruption of the fat globule, the surface tension rises to a high level of 15 mN/m and amphiphilic molecules in the plasma quickly adsorb to the lipid droplet to lower this value. The adsorbed layers consist mainly of serum proteins and casein micelles.

The following processes are of common interest in milk processing:

Coalescence is an irreversible increase in the size of fat globules and a loss of identity of the coalescing globules.

Flocculation is a reversible (with minor energy input) agglomeration/clustering of fat globules with no loss of identity of the globules in the floc. The fat globules that flocculate can be easily redispersed if they are held together by weak forces, or they might be harder to redisperse if they share part of their interfacial layers.

Partial coalescence is an irreversible agglomeration/clustering of fat globules, held together by a combination of fat crystals and liquid fat, and a retention of identity of individual globules as long as the crystal structure is maintained (i.e., temperature dependent, once the crystals melt, the cluster coalesces). They usually come together in a shear field, as in whipping, and it is envisioned that the crystals at the surface of the droplets are responsible for causing colliding globules to stick together, while the liquid fat partially flows between and acts as the cement. Partial coalescence dominates structure formation in whipped, aerated dairy emulsions, and it should be emphasized that crystals within the emulsion droplets are responsible for its occurrence.

10.2.1.2 Milk Lipids: Functional Properties

Like all fats, milk fat provides lubrication. They impart a creamy mouth feel as opposed to a dry texture. Fat globules produce a "shortening" effect in cheese by keeping the protein matrix extended to give a soft texture. Milk proteins are one of the most important constituents. The primary structure of proteins consists of a polypeptide chain of amino acids residues joined together by peptide linkages, which may also be cross-linked by disulfide bridges. Amino acids contain both a weak basic amino group and a weak acid carboxyl group connected to a hydrocarbon chain, which is unique to different amino acids. The three-dimensional organization of proteins, or conformation, also involves secondary, tertiary, and quaternary structures. The secondary structure refers to the spatial arrangement of amino acid residues that are near one another in the linear sequence. The alpha-helix and beta-pleated sheet are examples of secondary structures arising from regular and periodic steric relationships. The tertiary structure refers to the spatial arrangement of amino acid residues that are far apart in the linear sequence, giving rise to further coiling and folding. If the protein is tightly coiled and folded into a somewhat spherical shape, it is called a globular protein. If the protein consists of long polypeptide chains, which are intermolecularly linked, they are called fibrous proteins. There are some minor proteins that are associated with FGM.

The concentration of proteins in milk is as follows:

Protein	Grams/Liter	Percent of Total
Total protein	33	100
Caseins	26	80
Total whey proteins	6	19
Immunoglobulins	1	2

Caseins (as well as their structural form: casein micelles), whey proteins, and milk enzymes are important colloidal molecules in milk. The casein content of milk represents about 80% of milk proteins. Caseins exhibit solubility at pH 4.6. The common compositional factor is that caseins are conjugated proteins, most with phosphate group(s) esterified to serine residues. These phosphate groups are important to the structure of the casein micelle. Calcium binding by the individual caseins is proportional to the phosphate content. The conformation of caseins is much like that of denatured globular proteins. The high number of proline residues in caseins causes particular bending of the protein chain and inhibits the formation of close-packed, ordered, secondary structures. Caseins contain no disulfide bonds. As well, the lack of tertiary structure accounts for the stability of caseins against heat denaturation because there is very little structure to unfold. Without a tertiary structure, there is considerable exposure of hydrophobic residues. This results in strong association reactions of the caseins and renders them insoluble in water. Accordingly, strong amphiphilic protein acts like a detergent molecule. Self-association is temperature dependent, and will form a large polymer at 20°C but not at 4°C.

10.2.1.3 The Casein Micelle

As mentioned earlier, some specific molecules, such as soaps or detergents, aggregate to form micelles. Most, but not all, of the casein proteins exist in a colloidal particle known as the *casein micelle.* Its biological function is to carry large amounts of highly insoluble CaP to mammalian young in liquid form and to form a clot in the stomach for more efficient nutrition. In addition to casein proteins, calcium, and phosphate, the micelle also contains citrate, minor ions, lipase and plasmin enzymes, and entrapped milk serum. These micelles are rather porous structures, occupying about 4 ml/g and 6% to 12% of the total volume fraction of milk. The "casein submicelle" model has been prominent for the last several years, but there is not universal acceptance of this model and there is mounting research evidence to suggest that there is not a defined submicellar structure to the micelle at all. In the submicelle model, it is thought that there are small aggregates of whole casein, containing 10 to 100 casein molecules, called submicelles. It is thought that there are two different kinds of submicelles: with and without kappa-casein. These submicelles contain a hydrophobic core and are covered by a hydrophilic coat, which is at least partly comprised of the polar moieties of kappa-casein. The hydrophilic casein-macro-peptide (CMP) of the kappa-casein exists as a flexible hair. The open model also suggests there are more dense and less dense regions within the micelle, but there is less of a well-defined structure. In this model, calcium phosphate nanoclusters bind caseins and provide for the differences in density within the casein micelle.

Colloidal calcium phosphate (CCP) acts as a cement between the hundreds or even thousands of submicelles that form the casein micelle. Binding may be covalent or electrostatic.

The following factors must be considered when assessing the stability of the casein micelle.

- The role of Ca^{++} is very significant in milk. More than 90% of the calcium content of skim milk is associated in some way or another with the casein micelle. The removal of Ca++ leads to reversible dissociation of beta-casein without micellular disintegration. The addition of Ca^{++} leads to aggregation. The same occurs between the individual caseins in the micelle but not much because there is no secondary structure in casein proteins.
- Alpha and beta-caseins do not have any cysteine residues. If any S-S bonds occur within the micelle, they are not the driving force for stabilization. Caseins are among the most hydrophobic proteins and there is some evidence to suggest they play a role in the stability of the micelle. It must be remembered that hydrophobic interactions are very temperature sensitive.
- Electrostatic interactions—Some of the subunit interactions may be the result of ionic bonding, but the overall micellar structure is very loose and open.
- van der Waals Forces—There has been some success in relating these forces to micellar stability. However, the steric stabilization is also of some importance. Especially, the hairy layer interferes with interparticle approach. There are several factors that will affect the stability of the casein micelle system:
  - Salt content—Due to its affect on the calcium activity in the serum and calcium phosphate content of the micelles.
  - pH—Lowering the pH leads to dissolution of calcium phosphate until, at the isoelectric point (pH 4.6), all phosphate is dissolved and the caseins precipitate.
- Temperature—At around 4°C, beta-casein begins to dissociate from the micelle, at 0°C, there is no micellar aggregation; freezing produces a precipitate called cryocasein.

Casein micelle aggregation: Caseins are able to aggregate if the surface of the micelle is reactive. Although the casein micelle is fairly stable, there are four major ways in which aggregation can be induced:

1. Proteolytic enzymes, as in cheese manufacturing
2. Acid
3. Heat
4. Age gelation

During the secondary stage, the micelles aggregate. This is due to the loss of steric repulsion of the casein molecule as well as the loss of electrostatic repulsion due to the decrease in pH. As the pH approaches its isoelectric point (pH 4.6), the caseins aggregate (charge–charge repulsion disappears when molecules are with zero charge). The casein micelles also have a strong tendency to aggregate because of hydrophobic interactions. Calcium assists coagulation by creating isoelectric conditions and by acting as a bridge between micelles. The temperature at the time of coagulation is very important to both the primary and secondary stages. During the secondary stage, increased temperatures increase the hydrophobic reaction. The tertiary stage of coagulation involves the rearrangement of micelles after a gel has formed. There is a loss of paracasein identity as the milk curd firms and syneresis begins. It has been found that acidification causes the casein micelles to destabilize or aggregate by decreasing their electric charge to that of the isoelectric point. At the same time, the acidity of the medium increases the solubility of minerals so that organic calcium and phosphorus contained in the micelle gradually become soluble in the aqueous phase. Casein micelles disintegrate and casein precipitates. Aggregation occurs as a result of entropically driven hydrophobic interactions. At temperatures above the boiling point, casein micelles will irreversibly aggregate.

The most remarkable observation is that all 22 minerals considered to be essential to the human diet are present in milk. Some of these are sodium (Na), potassium (K), and chloride (Cl). The electroneutrality is maintained by free ions (negatively charged to lactose).

The viscosity of milk and milk products is reported to be important in the rate of creaming. The viscosity of milk increases with a decrease in temperature, because the increased voluminosity of casein micelle temperatures above 65°C increase viscosity due to the denaturation of whey proteins pH, an increase or decrease in pH of milk also causes an increase in casein micelle voluminosity. Fat globules that have undergone cold agglutination, may be dispersed due to agitation, causing a decrease in viscosity.

10.3 APPLICATIONS OF SCANNING PROBE MICROSCOPES TO SURFACE AND COLLOID CHEMISTRY

Microscopes have played an important role in science, which reveal the structures of the material. The degree of microscopic resolution determines the degree of information. Over many decades, the goal has been to be able to see single atoms or molecules. During the end of the 20th century, a big surge in the development of very important techniques has become available for science (nanoscience) and technology (self-assembly structures [micelles; monolayers, and vesicles], biomolecules, biosensors, surface and colloid chemistry, and nanotechnology). In fact, current literature shows that these developments have no end to this trend with regard to the vast expansion in the sensitivity and level of information.

Typical of all humans, seeing is believing, so the microscope has attracted much interest for many decades. All these inventions, of course, were basically initiated on the principles laid out by the telescope (as invented by Galileo) and the light-optical microscope (as invented by Hooke). Over the years, the magnification and the resolution of microscopes has improved. However, for the man to understand nature, the main aim has been to be able to see atoms or molecules. This goal has been achieved and the subject as described here will explain the latest developments that were invented only few decades ago.

The ultimate aim of scientists has always been to be able to see molecules while they are active. In order to achieve this goal, the microscope should be able to operate under ambient conditions. Further, all kinds of molecular interactions between a solid and its environment (gas or liquid or solid), initially, can take place only via the surface molecules of the interface. It is obvious that when a solid or liquid interacts with another phase, knowledge of the molecular structures at these interfaces is of interest. The term *surface* is generally used in the context of gas–liquid or gas–solid phase boundaries, whereas the term *interface* is used for liquid–liquid or liquid–solid phases. Furthermore, many fundamental properties of surfaces are characterized by morphology scales of the order of 1 to 20 nm (1 nm = 10^{-9} m = 10 Å [angstrom = 10^{-8} cm]).

Generally, the basic issues that should be addressed for these different interfaces are as follows:

- What do the molecules of a solid surface look like, and how are the characteristics of these different than the bulk molecules? In the case of crystals, one asks about the kinks and dislocations.
- Adsorption on solid surfaces requires the same information about the structure of the adsorbates and the adsorption site and configurations.
- Solid–adsorbate interaction energy is also required, as is known from the Hamaker theory.
- Molecular recognition in biological systems (active sites on the surfaces of macromolecule, antibody–antigen) and biological sensors (enzyme activity, biosensors).
- Self-assembly structures at interfaces.
- Semiconductors and applications.

Depending on the sensitivity and experimental conditions, the methods of molecular microscopy are many and varied. The applications of these microscopes are also very varied and extensive. For example, information about crystal structures as well as three-dimensional configurations of macromolecules has been obtained.

The most widely used application of microscopy is found in the case of surfaces and the study of molecules at the surfaces. Generally, the study of surfaces is dependent on understanding not only the reactivity of the surface but also the underlying structures that determine that reactivity. Understanding the effects of different morphologies may lead to a process for enhancement of a given morphology and hence to improved reaction selectivities and product yields.

Atoms or molecules at the surface of a solid have fewer neighbors as compared with atoms in the bulk phase, which is analogous to the liquid surface; therefore, the surface atoms are characterized by an unsaturated, bond-forming capability and accordingly are quite reactive. Until a decade ago, electron microscopy and some other similarly sensitive methods provided some information about the interfaces, although there were always some limitations inherent in all these techniques, which needed improvements.

A few decades ago, the best electron microscope images of globular proteins were virtually all but a little more than shapeless blobs. However, these days, due to relentless technical advances, electron crystallography is capable of producing images at resolutions close to those attained by x-ray crystallography or multidimensional nuclear magnetic resonance (NMR). To improve upon some of the limitations of the electron microscope, newer methods were needed. A decade ago, a new procedure for molecular microscopy was invented and will be delineated herein. The new scanning probe microscopes not only provided new kinds of information than hitherto as known from x-ray diffraction, for example, but these also opened up new areas of research (for example, nanoscience and nanotechnology).

The basic method for these scanning probe microscopes (SPMs) (Birdi, 2003) was essentially to be able to move a tip over the substrate surface with a sensor (probe) with molecular sensitivity (nm) in both the longitudinal and height direction (Figure 10.1). This may be compared with the act of sensing with a finger over a surface or more akin to the old-fashioned record player with a metallic needle (a probe for converting mechanical vibrations to music sound) on a vinyl record.

Scanning probe microscopy was invented by Binnig and Rohrer, who won the Nobel Prize in 1986 (Birdi, 2003, 2009). The scanning tunneling microscope (STM) was based upon scanning a probe (metallic tip) since it is a sharp tip just above the substrate while monitoring some interaction between the probe and the surface. The tip is controlled to within 0.1 Å (1 nm).

In SPM the various interactions between the tip and the substrate are as follows:

STM—The tunneling current between a metallic tip and a conducting substrate, which are very close in proximity but not actually touching in physical contact. This is controlled by piezo motors in a stepwise method.

AFM (atomic force microscope)—The tip is brought closer to the substrate while the van der Waals force is monitored. At a given force the piezo motor controls this setting while the surface is scanned in x–y direction.

FFM (friction force microscope)—This is a modification of AFM, where force is measured (Birdi, 2003).

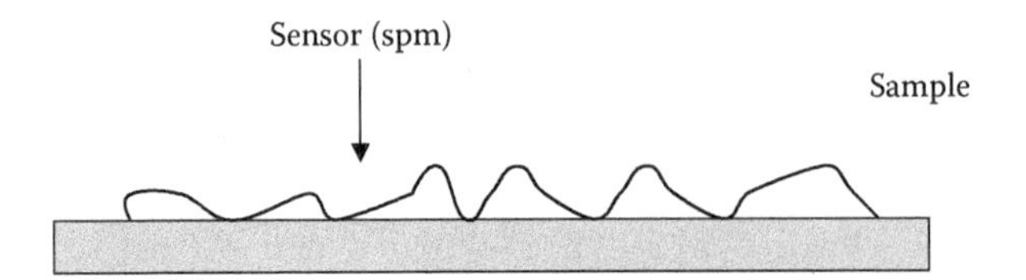

FIGURE 10.1 Scanning probe microscope (SPM). The probe can operate both in air or liquid media.

A schematic description of the SPM with the tip (dimension, 0.2 mm) and sample is shown in Figure 10.2.

The most significant difference between SPM and x-ray diffraction studies has been that former can be carried out both in the air and water (or any other fluid). Corrosion and similar systems have been investigated by using STM. The tip is covered by a plastic material and this allows one to operate STM under fluid environment. STM has been used to study the molecules adsorbed on solid surfaces. The Langmuir-Blodgett (L-B) films have been extensively investigated by both STM and AFM. AFM has been used to study surface molecules under different conditions.

AFM has allowed scientists to be able to study molecular forces between molecules at very small (almost molecular size) distances. Further, AFM is a very attractive and sensitive tool for such measurements. In a recent study the colloidal force as a function of pH of SiO_2 immersed in aqueous phase were reported using AFM. The force between an SiO_2 sphere (about 5 mm diameter) and a chromium oxide surface in aqueous phase of sodium phosphate were measured (pH from 3 to 11). The SiO_2 sphere was attached to the AFM sensor, as shown in Figure 10.3.

These data showed that the isoelectric point (IEP) of SiO_2 was around pH2, as expected. The binding of phosphate ions to chrome surface were also estimated as a

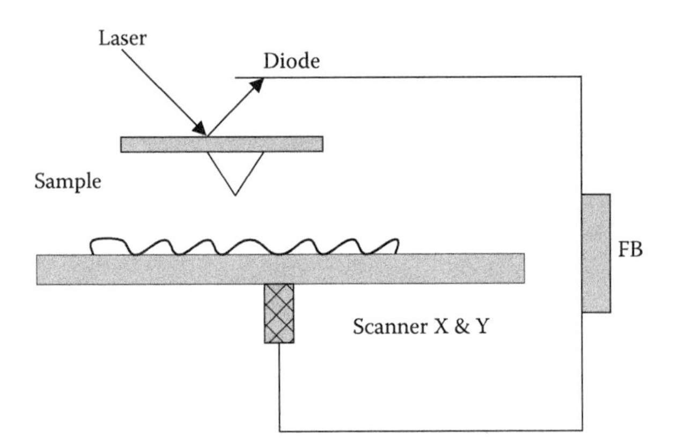

FIGURE 10.2 A schematic drawing of the sensor (tip, cantilever, optical, magnetic device) movement over a substrate in xyz direction with nanometer sensitivity (controlled by piezomotor) at solid–gas or solid–liquid interface.

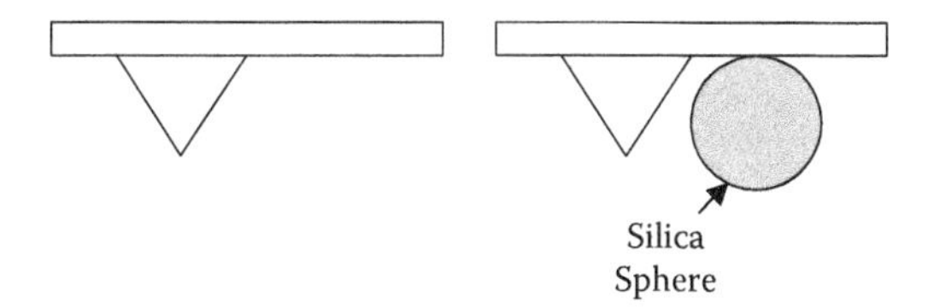

FIGURE 10.3 AFM sensor with an SiO_2 sphere.

function of pH and ionic strength (Birdi, 2003). Further, both STM and AFM have been used to investigate corrosion mechanisms of metals exposed to aqueous phase. Since both STM and AFM can operate underwater, this gives rise to a variety of possibilities. It has been found that the other setup can be used, where instead of SiO_2, other molecules can be used to investigate surfacc phenomena. For example, binding of bacteria to surfaces can be investigated by SPM methods. In the coming decades one expects many advances in the industrial applications of SPM.

10.3.1 Domain Patterns in Monomolecular Film Assemblies

Self-assembly monomolecular (SAMs) *films* at the air–water interface are very sensitive structures to any external forces. For example, if these monomolecular assemblies are compressed to certain values some drastic rearrangements will have to take place. Monolayers of lipids exhibit changes in structures near or after the collapse state, which has been designated as *domains*. The spontaneous formation of domain assemblies in monomolecular films of amphiphiles at air–water (oil–water interface needs) interface has evoked great interest. Considerable evidence seems to suggest that these two-dimensional assemblies arise in response to competing interactions. During the past decade an extensive number of investigations have been carried out by using different microscopes with varying sensitivities.

Molecular domains have been observed by using the AFM analyses of L-B films of collapsed state (Birdi, 1997, 2003). For example, cholesterol films showed half-butterfly-shaped domains (each domain consisting of 10^7 molecules) (Figure 10.4). This quantity was estimated from the following data. The height of domains was 90 Å, which corresponds with six layers of the cholesterol molecule (the length of molecule is found to be 15 Å from molecular models). The AFM image analysis is capable in calculating the area of the image. Since one knows the magnitude of the area/molecule of the cholesterol molecule (40 Å^2), this analysis can be carried out. Domains were found to be very regular in shape. In comparison to macrodomains, the latter domains thus are measured as three-dimensional by AFM. This indicates that such nanostructures can be investigated by using AFM. Thus in the future any nanosurface analysis will be useful in understanding these new technologies.

As compared to ordinary microscopes, the SPM provides 2D and 3D images. The 3D images allow one to see molecules with different diameters (Birdi, 2003). SPMs have contributed large, useful information about surface structures in the past decades (Table 10.1).

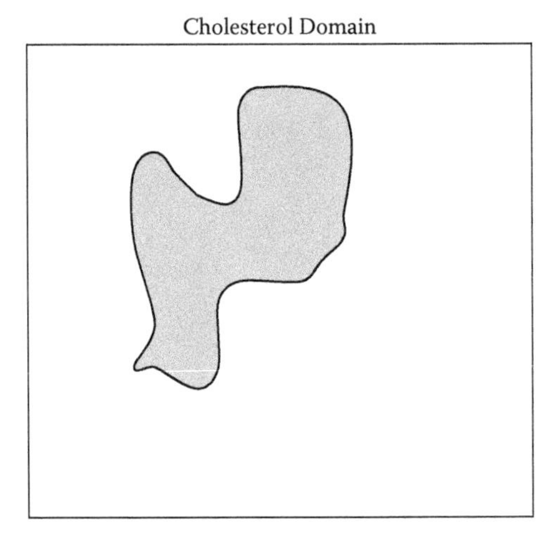

FIGURE 10.4 Cholesterol domain (a two-dimensional crystal) (3D schematic drawing consisting of 10^7 molecules) as studied by AFM (60000 × 60000 × 100 Å).

TABLE 10.1
Areas of Application for the STM and AFM

Areas of Application for the STM and AFM
Lipid monolayers (as Langmuir-Blodgett films)
Different layered substances on solids
Self-assembly structures at interfaces
Solid surfaces
Langmuir-Blodgett films
Thin-film technology
Interactions at surfaces of ion beams/laser damage
Nanoetching and lithography, nanotechnology
Semiconductors
Mineral surface morphology
Metal surfaces (roughness)
Microfabrication techniques
Optical and compact discs
Ceramic surface structures
Catalyses
Surface adsorption (metals, minerals)
Surface manipulation by STM/AFM
Polymers
Biopolymers (peptides, proteins, DNA, cells, virus)
Vaccines

10.4 DRUG DELIVERY DESIGN

In the medicinal industry one prescribes a sufficient dosage of a medicine, such that the medicine can be present in the blood with a particular and sufficient concentration. If one considers some examples, the range of concentration per dose (i.e., per tablet) may range from a microgram to 1000 mg. Since many of the medicinal molecules decompose in the body before reaching the target (such as liver, heart, and lungs), one uses a higher concentration (10 to 1000 times) than needed at the target. During the past decades, investigations have been reported on designing drug delivery techniques that target the medicine such that one can reduce the initial excess concentration.

10.5 BUILDING AND CEMENT INDUSTRY

Man has built houses using mudbricks and bricks and stones, and has used some form of cement to build houses and monuments for thousands of years. Some examples are the Egyptian pyramids (2000 BC), Roman Coliseum (2000 AD), and the Great Wall of China (200 BC).

This actually indicates that mankind has been aware of the role of surface and colloid chemistry (indirectly) for many centuries. In modern times, in addition to building small structures as houses, the very large constructions such as dams or high-rise buildings are most notable examples. The main aim of using cement is to bind two bricks. Cement consists of very fine particles and hardens after water has evaporated.

The chemical structures involved in cement are:

- Plaster of Paris ($CaSO_4 \cdot {}^1/_2\ H_2O$)

$$CaSO_4 \cdot {}^1/_2\ H_2O + 1{}^1/_2 H_2O = CaSO_4 \cdot 2H_2O \text{ (gypsum)}$$

- Lime-based cement (CaO)

$$CaO + H_2O = Ca(OH)_2 + CO_2\ CaCO_3 \text{ (calcite)}$$

The relatively high solubilities of portlandite ($Ca(OH)_2$) and gypsum means that they deteriorate rapidly in moist or wet environments. Many decades ago the Romans used lime-based cements and mortars (cement and sand) by ramming the wet pastes to form a high-density surface layer that carbonates in contact with air to produce a low permeability surface of calcite. Using this procedure leads to the protection of the $Ca(OH)_2$ layer (such as the Roman Hadrian's Wall). Lime mortars were still used in domestic construction until relatively recently. Hydration products are very insoluble cements set under water. It is likely that during the early Roman period silica and alumina and also volcanic earths were used as cements. Portland cement is made from finely ground limestone and finely divided clay to give a burned product containing 70% CaO, 20% SiO_2, 4% Fe_2O_3, and 4% Al_2O_3 plus smaller amounts of minor oxides (e.g., Na_2O, K_2O, and MgO). The major raw materials used for modern cement making are:

- Limestones and argillaceous shales
- Chalks
- Shells and clays
- Calcareous muds and other iron-bearing aluminosilicates

The essential step is the efficient grinding and blending of raw materials. The final properties of the cement strongly depend on its mineral composition so that raw composition and firing conditions are adjusted depending on the type of cement to be produced.

The interfacial zone consisted of:

- A thin (1 or 2 m thick) duplex film in actual contact with the reinforcement.
- Outside of this, a zone of perhaps 10 to 30 m thickness, which, in reasonably well-hydrated systems, is largely occupied by relatively massive calcium hydroxide crystals, with occasional interruptions of more porous regions.
- Outside of this, a highly porous layer parallel to the interface. The interaction of cracks initiated in the matrix with this interfacial zone was observed.

The hydration gives rise to effects on pore filling and consequent enhancement of mechanical performance (low porosity pastes are stronger than high-porosity ones).

The durability of cement pastes is strongly influenced by: (1) internal chemistry and (2) paste microstructure. Isolated pores are completely enclosed by hydration products so that material transport into and out the pore is limited. In the hardening stage the pore size distribution plays an important role. This arises from the fact that the curvature of liquid in different-sized pores varies due to the capillary forces. Connected porosity is that through which a continuous pathway between regions of the microstructure exists. Continuous or interconnected porosity often (although not always) links the interior of the paste to the outside world so that aggressive chemical species can penetrate and degrade the paste internally, having consequences for paste durability. The effect of the blending agents identified earlier on microstructure is to cause a reduction in the degree of interconnected porosity. This is especially true in the case of blast-furnace slag (BFS)-containing pastes. Although the overall porosity, as determined by neutron scattering is still significant, the interconnected porosity as measured by intrusion methods is low. Cement has been used in modern times to build houses and very large structures (such as dams). The cement industry is very large in volume and has an important impact on the social life of mankind. Especially considering the building of high-rises approaching the kilometer scale. The Hoover Dam has the following dimensions:

- Height, 250 m
- Weight, 6.6 million tons
- Volume of concrete used, 118 million ft^3
- Width, 250 m

It is thus obvious how important it is to have a knowledge of cement and its colloid structure for the implementation of such structures. It is important to consider that one of the greatest obstacles that builders had to overcome was a simple fact of concrete and cement chemistry. Concrete is made of cement, aggregate, and water mixed together to form a paste. The aggregate is usually a filler material composed of inert ingredients such as sand and rocks. When water is added, the components of cement undergo a chemical reaction known as hydration. As hydration occurs, the silicates are transformed into silicate hydrates and calcium hydroxide ($Ca(OH)2$), and the cement slowly forms a hardened paste. One of the products of the hydration reaction is heat, which is actually a great deal.

10.6 FRACKING INDUSTRY: GAS AND OIL RECOVERY FROM SHALE DEPOSITS

Gas and oil deposits around the world are being exploited for energy demands by mankind. However, the known reserves of gas and oil are being depleted faster than new reservoir discovery. It has been known for decades that gas is found in adsorbed state in shale reserves around the world (including Canada, the United States, Poland, China, Europe, United Kingdom, Russia, Australia, and India). Until recently recovery of adsorbed gas from shale was very costly as compared to other energy sources (EPA, 2013). Shale consists of layered deposits of riverbeds. The organic matter found in these layered structures over a million years has evolved into mostly methane gas. Gas is very tightly held within the shale structure and not easily available. In the gas and oil recovery from shale one bores down to the reservoir and pumps water with sand and additives. In this process small explosions are created to make shale release gas/oil for recovery. The process is specific for each shale reservoir type.

Shale fracking technology:

```
Layer in shale = = = = = = = = = = = = = = = = = = = = = =
= = = = = = = = = =
= = = = = = = = = = = = = = = = = = = = = = = = = = = = = =
= = =
= = = = = = = = = = = = = = = = = = = = = = = = = = = = = =
= = =
= = = = = = GAS = = = = = GAS = = = = = = = GAS = = = = = =
= = = = = = = = = = = = = = = = = = = = = = = = = = = = = =
= = =
= = = = = = = = = = = = = = = = = = = = = = = = = = = = = =
= = =
```

The layered salt deposits contain tightly held gas or oil in shale deposits. The recovery of gas is obtained by creating fractures (by inducing controlled explosions) in the reservoir thus releasing gas for recovery. The trapped gas is released and recovered for energy production.

10.7 OTHER INDUSTRIAL APPLICATIONS

10.7.1 Paper Industry (Surface and Colloid Aspects)

The paper industry today is the most advanced industrial innovation. Not only can the thickness of the paper for printing (such as 80, 90, or 100 g/cm^2) be specified but the paper gloss or paper brightness, as well. In fact, in many quantitative analyses (such as measurement of area under a curve) one uses the weight of the paper. Since the weight is very reliable, such analyses are routinely used (such as curve areas). These properties are much dependent on the surface chemical properties of the paper. During paper manufacturing, colloidal chemistry is applied to control the implications of additives (such as fillers, sizing, dyes, and strength additives). The paper acquires a negative charge due to oxidation when exposed to air (mostly -COO- charges, with $pK_a = 5$). This suggests that at pH = 5, the paper pulp will have an almost zero charge.

The paper industry deals with various substances, including:

- Cellulose
- Lignin
- Dyes
- Dispersants
- Fillers ($CaCO_3$, TiO_2 [whitener])
- Alum
- Defoamers

10.7.2 Inks and Printers (Colloid Chemistry)

Mankind has been applying inks for writing for thousands of years. The ancient people used naturally occurring colloidal fine articles from ashes (mostly charcoal) dispersed in oil (olive oil). Modern ink-jet printers employing color are based upon much more sophisticated systems. These ink-jet printers have a number of nozzles that inject ink droplets at the surface of paper. Simultaneously, different colors (from five to seven different color bottles) are mixed to obtain the desired color shade (more than hundreds of thousands). In a typical printer there may be 30,000 injections per second and there may be more than 500 nozzles (each with the size less than a human hair [$\mu m = 10^{-6}$ m]). Further, the ink is found to have shelf life of more than a year. In this process, the surface and colloid principles of most importance are:

- Contact angle (ink and the nozzle, ink and the surface of the paper)
- Capillary pressure (at the nozzle)

It is thus seen that surface and colloid principles are involved in a much complex manner in these printers.

10.7.3 Theory of Adhesives and Adhesion

Adhesives are used in everyday applications. Adhesives may be in liquid form or thick pastes. The main mechanism is based upon the polymerization or cross-linking of polymers, which gives rise to glue or other adhesive applications. The degree of adhesion of such a process is determined by conventional technological tests.

And what follows from Young's equation (Chapter 5, Equation 5.1)—if a liquid is removed from the surface of a solid, there will be work needed, W_{ad}, per square cm of solid exposed.

This process will require the destruction of 1 cm^2 of interface γ_{SL}, and the creation of 1 cm^2 of γ_S and γ_L. From this we get:

$$W_{ad} = \gamma_S + \gamma_L - \gamma_{SL} \quad (10.1)$$

which on combining with Young's equation gives:

$$W_{ad} = \gamma_L (1 + \cos \theta) \quad (10.2)$$

This shows that to remove a liquid from a solid surface the process is dependent on surface tension and contact angle, θ. If a liquid wets the solid surfaces (water on glass $\theta = 0$),

$$W_{ad} = 2 \gamma_L \quad (10.3)$$

W_{ad} is thus the work needed to create twice the surface tension (one on each side after separation or breakup).

Another area of much interest is the adhesion of ice to solids. Especially in countries such as Greenland, the movement of vehicles requires special knowledge about the adhesion of various materials to ice. This system is obviously of much interest in general everyday phenomena, for example, tire friction on road surfaces, ice on metal surfaces, and ships, especially the adhesion of ice on ships sailing in the cold areas and on wings of airplanes. Investigations have shown that the adhesive bonds between clean metal surface and ice are very strong. When the ice is removed by force, it breaks while leaving a thin layer of ice on solid. Further, one experiences a large attraction between two smooth solid surfaces and a drop of liquid (say, water and two glass plates). It is thus obvious that the peeling of two plates in this system will increase with a glue or something similar (instead of water).

Surface coatings of solids: The properties of solid surface can be changed while the rest of the material remains unchanged. These coatings are numerous and new applications are developed continuously.

The most recent is called Sol-Gel (solution and gelling) coating. The principle in this coating process is as follows:

- Solid surface is coated with a metal alkyl-oxide solution.
- The metal alkyl-oxide is hydrolyzed and forms a gel.
- The gel forms a network on the solid surface.
- The gel is exposed to high temperatures leaving behind a transparent coating.

The alkyl oxide, $(Si\ (RO)_4)n$, is converted by a catalyst and after dealkylation it forms a polymer of $(Si(OH)_4)n$. Upon heating and dehydration, one gets a coating of $(Si(O)_4)n$. The thickness of coating on aluminum (or other metal surface) can be 3 to 10 μm.

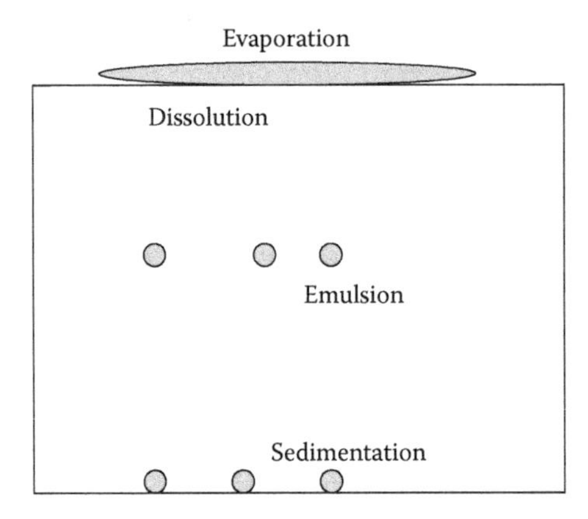

FIGURE 10.5 An oil spill on the ocean and its treatment mechanisms (evaporation, emulsion, dissolution, sedimentation).

10.7.4 Oil Spills and Cleanup Processes on Oceans

The worldwide concern with oil spills and their treatment is very dependent on surface chemistry principles. Oil is transported across the seas in very large amounts (over 80 million barrels per day). Oil spills on sea surfaces (Figure 10.5) are exposed to various parameters:

- Loss by evaporation
- Loss by sinking to the bottom (as such or in conjunction with solids)
- Emulsification.

Oil spills are treated by various methods, depending on the region (in warmer seas or around cold climate). It is also apparent that an oil spill in the Gulf of Mexico will be of completely different nature than a similar accident near Greenland, for instance. The composition of oil differs from place to place. The light fluids of oil will evaporate into the air. The oil that has been adsorbed on solid suspension will sink to the bottom. The remaining oil is sometimes skimmed off by suitable machines. In some cases, surfactants are used to emulsify the oil and this emulsion slowly sinks to the bottom. However, no two oil spills are the same because of the variation in oil types, locations, and weather conditions involved. Although broadly speaking, there are four main methods of response.

1. Leave the oil alone so that it breaks down by natural means. If there is no possibility of the oil polluting coastal regions or marine industries, the best method is to leave it to disperse by natural means. A combination of wind, sun, current, and wave action will rapidly disperse and evaporate most oils. Light oils will disperse more quickly than heavy oils. The temperature of the seawater will also have an effect on the evaporation process.
2. Contain the spill with booms and collect it from the water surface using skimmer equipment. Spilled oil floats on water and initially forms a slick

that is a few millimeters thick. There are various types of booms that can be used either to surround and isolate a slick, or to block the passage of a slick to vulnerable areas such as the intake of a desalination plant or fish-farm pens or other sensitive locations. Boom types vary from inflatable neoprene tubes to solid but buoyant material. Most rise up to about a meter above the waterline. Some are designed to sit flush on tidal flats while others are applicable to deeper water and have skirts that hang down about a meter below the waterline. Skimmers float across the top of the slick contained within the boom and suck or scoop the oil into storage tanks on nearby vessels or on the shore. However, booms and skimmers are less effective when deployed in high winds and high seas.

3. Use dispersants to break up the oil and speed its natural biodegradation. Dispersants act by reducing the surface tension that inhibits oil and water from mixing. Small droplets of oil are then formed, which helps promote rapid dilution of the oil by water movements. The formation of droplets also increases the oil surface area, thus increasing the exposure to natural evaporation and bacterial action. Dispersants are most effective when used within an hour or two of the initial spill. However, they are not appropriate for all oils and all locations. Successful dispersion of oil through the water column can affect marine organisms like deep-water corals and sea grass. It can also cause oil to be temporarily accumulated by subtidal seafood. Decisions on whether to use dispersants to combat an oil spill must be made in each individual case. The decision will take into account the time since the spill, the weather conditions, the particular environment involved, and the type of oil that has been spilled.
4. Introduce biological agents to the spill to hasten biodegradation. Most of the components of oil washed up along a shoreline can be broken down by bacteria and other microorganisms into harmless substances such as fatty acids and carbon dioxide. This action is called biodegradation.

The oil spill technology is very advanced and can operate under very divergent conditions (from the cold sea waters to the tropic seas). The biggest difference arises from oils that may contain varying amounts of heavy components (such as tar).

Emulsion polymerization: Emulsion polymerization is one of the major examples where detergents are applied to create microreactions. For instance, to polymerize styrene (which is insoluble in water), one adds an initiator to the aqueous phase. The polymer, polystyrene (PS), is formed and the suspension is stabilized by using suitable emulsifiers. The latex thus formed is used in various industrial applications.

For example PS can be prepared as follows. A mixture of styrene, detergent (Na-dodecanoate), and water is agitated by an ultrasonic mixer to produce a fine emulsion. On the addition of hydrogen peroxide (initiator), one obtains PS as a polymer. PS can be extracted after filtration. The polymer molecular weight is determined by various methods (such as light scattering or osmotic pressure).

Polymer colloid systems: Polymers are large molecules, with molecular weights varying from over a thousand to millions. Polymers play a important role in both biological and industrial systems. Polymers may exhibit neutral or negative or positive charges. Recently there has been much interest in developing methods to apply polymer colloids. The principle is to employ polymer colloidal beads instead of using large solid particles (Birdi, 2009). These colloids are dispersed in aqueous media. These colloids exhibit two-fold progenies: polymers and colloidal.

10.7.5 Photographic Industry (Emulsion Films)

Photographic film production has been a very sophisticated technical innovation, even though this process is now more taken over by the digital camera technique. In general, photographic film consists of a sheet of plastic base (polyester or nitrocellulose [celluloid]), which is coated with a very thin emulsion-containing light-sensitive Ag salts. The latter is bonded to gelatin gel. The film is exposed to image created by the lens. The resolution of the image produced is determined by the particle size of the Ag particles. Most of the industry is prone to high secrecy (patents) and not much is known how the different producers of film achieve the end result. Black-and-white films were much simpler than color films. Further, the chemical reactions that take place on the surface of the film may vary from minutes (or longer) to merely 1/1000 second. This indeed demands much technology know-how to meet these demands.

10.7.6 Fire Fighting and Diverse Applications of Foams

Even though the foam formation is a simple process, that is, blowing air into a solution of surface active agent in water, its applications are far too involved, especially in those cases where foam is the main characteristic of the end product. Some useful examples are discussed next.

In fire fighting, water is the most common procedure. However, alone water in many cases is not efficient due to some limitations. The main solution is to contain the fire by covering the flames with a layer of foam such that contact with air (i.e., oxygen) is prevented. In the case of application of foams in fire fighting, the rate of drainage needs to be considered. The foam fire-fighting procedure is useful since it gives slower evaporation and thus a more useful result. The thick foam also helps in reducing the vapors (of organic fluids).

In foam rubber, foamed polymers, shaving foams, milk shakes, and whipped creams, one needs slowly draining thin liquid films. Accordingly, in such industrial foam applications the rate of drainage is the most important factor.

10.7.7 Coal Slurry Applications

Mankind's need for fuel and energy is becoming increasingly critical, as far as resources and dependability are concerned. The largest known energy resource is

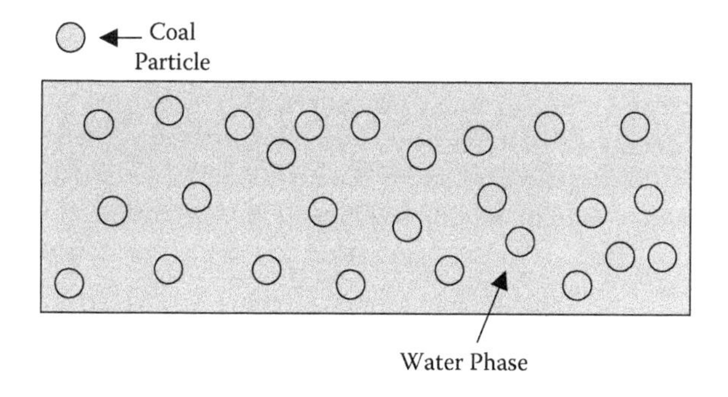

FIGURE 10.6 Coal slurry transport via pipelines.

coal, with supplies lasting for over 1000 years. The oil and gas supply is in the range of less than 100 years at the current rate of use of approximately 90 million barrels per day. The coal reserves are very large, but the mobility of coal is difficult (by truck, train, or ship loads) as compared to the oil and gas (by pipelines). In order to transfer coal in pipelines, the making of *coal slurries* is initiated (Figure 10.6). Coal is finely divided and after that it is dispersed in water or oil such that a suitable slurry for pipeline transfer can be suitably achieved (Papachristodoulou and Trass, 1987) (Figure 10.6). Something similar is already being developed in the shale oil industry (in Canada). Coal can thus be transported through pipelines after being dispersed in conjunction with water–oil. This industrial development is based on many aspects related to the surface properties of coal. Coal is associated with the high CO_2 production in the burning plants. Recently, however, great advances have been made in CO_2 capture. This process deals with the passing of gases from the coal power plants through a solution to convert CO_2 to carbonates. The average concentration of CO_2 in air is approximately 400 ppm by volume. Further, concentration of CO_2 falls during the summer as plants consume it through the photosynthesis cycle. There is about 50% as much CO_2 dissolved in the oceans (it is surface process), than in atmosphere. Carbon capture has been carried out in some industrial processes, as follows:

- CO_2 is separated from exhausts and stored in secure underground formations.
- CO_2 is used in, for example, enhanced oil recovery, cement and building.
- CO_2 is pumped deep into oceans and stored as carbonate salts.

In recent decades, there has been a very large interest in this technology. At this point, about 23 million tons of CO_2 is being captured. The research is based on studying the surface properties of coal and its adsorption characteristics. The oil production from shales is increasing as the oil prices increase. Additionally, in many coal operations the dust is controlled by wash water procedure.

Coal-water slurry fuel (CWS or CWSF or CWF) is a fuel that consists of fine coal particles suspended in water. The presence of water in CWS reduces harmful emissions into the atmosphere, makes the coal explosion-proof, makes use of coal equivalent to use of liquid fuel (e.g., heating oil), and gives other benefits. CWS consists of 55% to 70% of fine dispersed coal particles and 30% to 45% water. Coal particles suitable for diesel fuel replacement are typically less than 20 μm in diameter. CWS can be used in place of oil and gas in any size of heating and power station. CWS is suitable for existing gas, oil, and coal boilers. During the last 30 years the U.S. Department of Energy (DOE) has been researching the use of coal water fuels in boilers, gas turbines, and diesel engines. When used in low-speed diesels, CWS has a thermal efficiency rating that rivals combined cycle gas turbines that burn natural gas as their primary fuel. It has been suggested that slightly modified modular diesel engine power plants that burn CWS are economically competitive with natural gas fired peaking electric plants in the 10 MWe to 100 MWe range of power supply. Because of the relatively low cost of coal when compared to other energy sources, CWS is a very competitive alternative to heating oil and gas. Depending on the geographical area the price per unit energy of CWS may be 30% to 70% lower than the equivalent oil or gas. The quality of coal, such as that with low BTU, has been found to make CWS a very cost-effective and environmentally friendly fuel for heat and power generation. CWS has also been used directly for combustion in steam boilers. In addition to this, there are other useful properties for using this method. The CWS production process also helps in the separation of noncarbon material mixed with the coal before treatment. This gives rise to a reduction of ash content to as low as 2%. Ball mills are used to pulverize and mix the coal sludge with water, producing a coal-water slurry. The CWS process consists of three major stages:

1. Coal is crushed to 10 to 65 μm particles.
2. The fine coal powder is treated in the wet milling and homogenization step, and mixing with water and some additives.
3. Coal is finely crushed to fine powder. Coal powder is mixed with water and additives to obtain a stable slurry. Coal slurry can be transported by suitable pipelines.

Different kinds of coal have been used in these applications. The CWS technology combined with carbon capture could be a useful process which could reduce CO_2 pollution and provide energy for many hundreds of years (*Global CCS Institute Newsletter*, 2013).

References

Adam, N. K., *The Physics and Chemistry of Surfaces*, Clarendon Press, Oxford, 1930.

Adam, N. K., *Physics and Chemistry of Surfaces*, Oxford University Press, Oxford, 1941.

Adamson, A. W., and Gast, A. P., *Physical Chemistry of Surfaces*, 6th ed., Wiley-Interscience, New York, 1997.

Agarwal, J. K., Breakdown in ultra-thin Langmuir films, *Thin Solid Films*, 27, 49, 1974.

Aveyard, R., and Hayden, D. A., *An Introduction to Principles of Surface Chemistry*, Cambridge, London, 1973.

Avnir, D., Editor, *The Fractal Approach to Heterogeneous Chemistry*, Wiley, New York, 1989.

Bakker, G., *Kapillaritat und Uberflachenspannung Handbuch der Eksperimentalphysik*, 3rd ed., Leipzig, 1928.

Bancroft, W. D., *Applied Colloid Chemistry*, McGraw-Hill, New York, 1932.

Barnes, G., *Interfacial Science (An Introduction)*, Oxford University Press, Oxford, 2011.

Becher, P., *Emulsions Theory and Practice,* 3rd ed., Oxford University Press, New York, 2001.

Birdi, K. S., Contact angle hysteresis, *J. Colloid Polym. Sci.*, 260, 8, 1982.

Birdi, K. S., *Lipid and Biopolymer Monolayers at Liquid Interfaces*, Plenum Press, New York, 1989.

Birdi, K. S., *Self-Assembly Monolayer (SAM) Structures*, Plenum Press, New York, 1999.

Birdi, K. S., *Scanning Probe Microscopes (SPM)*, CRC Press, Boca Raton, FL, 2003.

Birdi, K. S., *Surface & Colloid Chemistry*, CRC Press, Boca Raton, FL, 2010a.

Birdi, K. S., *Fractals (In Chemistry, Geochemistry and Biophysics)*, Plenum Press, New York, 1993.

Birdi, K. S., Editor, *Handbook of Surface & Colloid Chemistry*, CRC Press, Boca Raton, FL, 1997.

Birdi, K. S., Editor, *Handbook of Surface & Colloid Chemistry—CD Rom*, 2nd ed., CRC Press, Boca Raton, FL, 2002.

Birdi, K. S., Editor, *Handbook of Surface & Colloid Chemistry*, 3rd ed., CRC Press, Boca Raton, FL, 2009.

Birdi, K.S., Editor, *Interfacial Electrical Phenomena*, CRC Press, Boca Raton, FL, 2010b.

Birdi, K. S., Vu, D. T., Winter, A., and Naargard, A., Capillary rise of liquids in rectangular tubings, *J. Colloid Polym. Sci.*, 266, 5, 1988.

Biresaw, G., and Mittal, K. L., *Surfactants in Tribology*, CRC Press, New York, 2008.

Boys, C. V., *Soap Bubbles*, Dover, New York, 1959.

Chattoraj, D. K., and Birdi, K. S., *Adsorption and the Gibbs Surface Excess*, Plenum Press, New York, 1984.

Cini, R., Loglio, G., and Ficalbi, A., Surface tension of water, *J. Colloid Interface Sci.*, 41, 287, 1972.

Davies, J. T., and Rideal, E. K., *Interfacial Phenomena*, Academic Press, New York, 1963.

Defay, R., Prigogine, I., Bellemans, A., and Everett, D. H., *Surface Tension and Adsorption*, Longmans, Green, London, 1966.

Diebold, U., Surface science of TiO_2, *Surface Science Reports*, 48, 53, 2003.

Dukhin, A. S., and Goetz, P. J., *Ultrasound Characterizing Colloids*, Elsevier, New York, 2002.

Environmental Protection Agency (EPA), and U.S. Geological Survey, *Complete Guide to Hydraulic Fracturing for Shale Oil and Natural Gas*, 2013.

Fanum, M., *Microemulsions*, CRC Press, New York, 2008.

Feder, J., *Fractals: Physics of Solids and Liquids*, Plenum Press, New York, 1988.

Fendler, J. H., and Fendler, E. J., *Catalysis in Micellar and Macromolecular Systems*, Academic Press, New York, 1975.

Friberg, S., Larsson, K., and Sjoblom, J., *Food Emulsions*, CRC Press, Boca Raton, FL, 2003.
Fuerstenau, M. C., Miller, J. D., and Kuhn, M. C., *Chemistry of Flotation*, Society of Mining Engineers, New York, 1985.
Gaines, G. L., Jr., *Insoluble Monolayers at Liquid-Gas Interfaces*, Wiley-Interscience, New York, 1968.
Gisler, T., Schultz, S. F., Borkovec, M., Sticher, H., Schurtenberger, P., D'Aguanno, B., and Klein, R., Colloidal charge renormalization, *J. Chem. Phys.*, 101, 1, 1994.
Gitis, N., and Sivamani, R., *Tribology Transactions*, Taylor & Francis, New York, 2004.
Global CCS (Carbon Capture and Storage) Institute Newsletter, April 2013.
Goodrich, F. C., Rusanov, A. I., and Sonntag, H., *The Modern Theory of Capillarity*, Akademie Verlag, Berlin, 1981.
Grodzka, J., and Pomianowski, A., DLVO theory, *Physicochem. Prob. Mineral Process.*, 39, 11, 2005.
Hansen, M. C., *Hansen Solubility Parameters*, Taylor & Francis, New York, 2007.
Harkins, W. D., *The Physical Chemistry of Surface Films*, Reinhold, New York, 1952.
Jaycock, M. K., and Parfitt, G. D., *Chemistry of Interfaces*, John Wiley & Sons, New York, 1981.
Kamperman, M., and Synytska, A., Switchable adhesion, *J. Mater. Chem.*, 22, 19390, 2012.
Koch, J. P., Fractal geometry in biopharmaceutical sciences, *Pharmazie*, 48, 643, 1993.
Kolasinski, K. W., *Surface Science*, John Wiley & Sons, New York, 2008.
Kristensen, D., Milk–fat globule structures, Ph.D. thesis, School of Pharmacy, Copenhagen, 1997.
Krungleviclute, V., Migone, A. D., Yudasaka, M., Ijima, S., CO_2 adsorption on carbon surface, *J. Phys. Chem., C*, 116, 306, 2012.
Kuespert, D., and Donohue, M. D., *J. Phys. Chem.*, 99, 4805, 1995.
Lara, J., Blunt, T., Kotvis, P., Riga, A., and Tysoe, W. T., Fractal geometry in biopharmaceutical sciences, *J. Phys. Chem., B*, 102, 1703, 1998.
Lovett, D., *Demonstrating Science with Soap Films*, Institute of Physics, Bristol, U.K., 1994.
Lyklema, J., *Fundamentals of Interfaces and Colloid Science, III, Liquid-Fluid Interfaces*, Academic Press, San Diego, 2000.
Matijevic, E., Editor, *Surface and Colloid Science*, Vol. 1, Wiley-Interscience, New York, 1969.
Matijevic, E., Editor, *Surface and Colloid Science*, Vol. 9, Wiley-Interscience, New York, 1976.
Miller, C. A., and Neogi, P., *Interfacial Phenomena*, CRC Press, New York, 2008.
Mousny, M., Omelon, S., Wise, L., Everett, E. T., Dumitriu, M., and Holmyar, D. P., Fluoride effects on bone formation, *Bone*, 43, 1067, 2008.
Neumann, A. W., *Applied Surface Thermodynamics*, CRC Press, New York, 2010.
Papachristodoulou, G., and Trass, O., Coal slurry technology, *Can. J. Chem. Eng.*, 65, 177, 1987.
Partington, J. R., *An Advanced Treatise of Physical Chemistry*, Vol. II, Longmans, Green, New York, 1951.
Rao, J. P., and Geckeler, K. E., Polymer nano-particles, *Progress Polymer Science*, 36, 887, 2011.
Reimhult, E., Hook, F., and Kasemo, B., Vesicle adsorption on silica, *Langmuir*, 19, 1681, 2003.
Rosen, M. J., *Surfactants and Interfacial Phenomena*, Wiley-Interscience, New York, 2004.
Ruiz, C. C., *Sugar-Based Surfactants*, CRC Press, New York, 2008.
Schramm, L. L., *Emulsions, Foams and Suspensions*, John Wiley, New York, 2005.
Sjoblom, J., in *Handbook of Surface and Colloid Chemistry*, 3rd ed., K.S. Birdi, Editor, CRC Press, Boca Raton, FL, 2008.
Soltis, A. N., Chen, J., Atkin, L. Q., and Hendy, S., Magnetic force spectroscopy of solid surfaces, *Curr. Appl. Phys.*, 4, 152, 2004.
Somarajai, G. A., and Li, Y., *Introduction to Surface Chemistry and Catalysis*, John Wiley & Sons, New York, 2010.
Somasundaran, P., *Colloidal and Surfactant Sciences*, CRC Press, New York, 2006.
Stefan, J., Surface tension and enthalpy of vaporization of liquids, *Ann. Phys.*, 29, 655, 1886.

Tanford, C., *The Hydrophobic Effect*, John Wiley, New York, 1980.
Taylor, R. P., Virtual fractals, *Nonlinear Dynam. Phys. Life Sci.*, 15, 129, 2011.
Woodson, M., and Liu, C., Electrodeposition for energy applications, *Phys. Chem. Phys.*, 9, 207, 2007.
Wu, X. Z., Ocko, B. M., Sirota, C. B., Sinha, S. K., Deutsch, M., Cao, B. H., and Kim, M. W., A molecular dynamics simulation of the surface-ordered phase of alkanes, *Science*, 261, 1018, 1993.
Zana, R., *Giant Micelles*, CRC Press, New York, 2008.
Zhuravlev, L. T., Surface chemistry of silica, *Colloids Surf.*, 1, 173, 2000.
Zoller, U., *Sustainable Development of Detergents*, CRC Press, New York, 2008.

Appendix: Common Fundamental Constants

Angstrom, Å = 10^{-8} cm = 10^{-10}m
Nanometer, nm = 10^{-9} m
Micrometer, μm = 10^{-6} m
1 Joule (J) = 10^{-7} erg = 0.239 cal = 6.242 10^{10} eV
1 kJ mol^{-1} = 0.239 kcal mol^{-1} = 0.404 k T per molecule (at 298 K)
1 N = 10^5 dyne
g = 9.81 m s^{-2}
Boltzmann constant, k_B = 1.381 × 10^{-23} J K^{-1}
Electronic charge, e = 1.602 × 10^{-19} C
Ideal gas law, P V = k T = 4.12 × 10^{-22} J at 298 K (25°C)
Ideal two-dimensional monolayer, ∏ A = k T = 411 (∏ = mN/m; A = A)
1 cal = 4.184 J
1 erg = 10^{-7} J
1 Atm = 1.013 × 10^5 N m^{-2} (Pa)
1 Pascal (Pa) = 1 N m^{-2} = 10 dyne/cm = 9.872 10^{-6} atm = 7.5 10^{-3} torr = 1.45 10^{-4} psi
1 bar = 10^5 N m^{-2} = 10^{-5} Pa = 0.987 atm = 750 mm Hg
kT/e = 25.7 mV (at 25°C)
Permittivity of free space, eo = 8.854 × 10^{-12} C^2 J^{-1} m^{-1}
eV = 1.6021 10^{-19} J
Kilowatt hour, kWh = 3.6 10^6 J
Viscosity of water, 0.001 N s m^{-2} = 1 cps, at 20°C
Dielectric constant of water, 80.2 at 20°C
Surface tension of water at different temperatures:
γ = 75.69 – 0.1413 T – 0.0002985 T^2
= T (where T is in centigrade)
(at 25°C = 71.97 mN/m)

Index

V

W

Y

Z